2023 台达杯国际太阳能建筑设计竞赛获奖作品集

Awarded Works from International Solar Building Design Competition 2023

阳光·零碳建筑
SUNSHINE & ZERO-CARBON ARCHITECTURE

中国建筑设计研究院有限公司　编
Edited by China Architecture Design & Research Group

执行主编：张　磊　鞠晓磊　张星儿
Chief Editor: Zhang Lei　Ju Xiaolei　Zhang Xinger

中国建筑工业出版社
CHINA ARCHITECTURE & BUILDING PRESS

图书在版编目（CIP）数据

阳光·零碳建筑：2023台达杯国际太阳能建筑设计竞赛获奖作品集=SUNSHINE & ZERO-CARBON ARCHITECTURE Awarded Works from International Solar Building Design Competition 2023 / 中国建筑设计研究院有限公司编；张磊，鞠晓磊，张星儿执行主编 . -- 北京：中国建筑工业出版社，2024.6. -- ISBN 978-7-112-30073-0

Ⅰ. TU18

中国国家版本馆CIP数据核字第2024FM7477号

为积极响应"双碳"目标，推动校园建筑低碳化，不断提升建筑品质，2023台达杯国际太阳能建筑设计竞赛以"阳光·零碳建筑"为主题，向全球征集作品。竞赛设置两个赛题，分别为零碳设计项目"广州科教城文化科技馆"和零碳提升项目"广州市公用事业技师学院社团综合楼"。该作品集收录了参赛项目中的一、二、三等奖和优秀奖作品，旨在推动可再生能源在建筑上的应用，打造科教城绿色低碳理念的展示窗口，探索零碳建筑的建设路径。本书适用于从事建筑学各分支领域研究的研究人员、工程技术人员、科技管理人员和高等院校师生阅读参考。

责任编辑：唐　旭　吴　绫　张　华
责任校对：张惠雯

阳光·零碳建筑
SUNSHINE & ZERO-CARBON ARCHITECTURE
2023台达杯国际太阳能建筑设计竞赛获奖作品集
Awarded Works from International Solar Building Design Competition 2023
中国建筑设计研究院有限公司　编
Edited by China Architecture Design & Research Group
执行主编：张　磊　鞠晓磊　张星儿
Chief Editor: Zhang Lei　Ju Xiaolei　Zhang Xinger

*

中国建筑工业出版社出版、发行（北京海淀三里河路9号）
各地新华书店、建筑书店经销
北京雅盈中佳图文设计公司制版
临西县阅读时光印刷有限公司印刷

*

开本：787毫米×1092毫米　1/12　印张：25　字数：752千字
2024年6月第一版　2024年6月第一次印刷
定价：178.00元
ISBN 978-7-112-30073-0
（43186）

版权所有　翻印必究
如有内容及印装质量问题，请与本社读者服务中心联系
电话：（010）58337283　QQ：2885381756
（地址：北京海淀三里河路9号中国建筑工业出版社604室　邮政编码：100037）

随着工业化和城市化的快速发展，碳排放量不断攀升，给地球生态环境带来了前所未有的压力，降碳已成为全球关注的热点。党的二十大报告中指出，推动绿色低碳高质量发展，实现碳达峰、碳中和是一场广泛而深刻的经济社会系统性变革，需要社会各领域统筹协调，各行业协同发力。本届竞赛设置广州科教城文化科技馆零碳设计和广州市公用事业技师学院社团综合楼零碳提升两项赛题，探索校园建筑的零碳实施路径。参赛者们通过作品展现出可喜的创新力与专业素养，承担起时代赋予的责任和使命，成为生态文明建设的设计者、建设者和传承者。

感谢台达集团资助举办2023台达杯国际太阳能建筑设计竞赛！

谨以本书献给致力于生态文明建设与绿色低碳发展的同仁们！

With the rapid advancement of industrialization and urbanization, carbon emissions continue to escalate dramatically, exerting unprecedented pressure to the earth's ecological environment. Consequently, carbon reduction has emerged as a global focus. As addressed in the report at the 20th National Congress of CPC, the promotion of green, low-carbon and high-quality development along with achieving carbon peak and carbon neutrality brings forth a systematic transformation with extensive and profound impact on economy and society. This transformation demands consolidated coordination from all walks of life and concerted efforts from various industries.

To embody these ideas and visions, Competition 2023 has set up two tasks: 1. Zero-carbon design for the Cultural and Scientific-technological Museum of Guangzhou Science and Education Park; and 2. Zero-carbon promotion for the Complex Building for Students Societies of Guangzhou Public Utility Technician College. Both tasks aim to explore a path towards zero-carbon implementation on campus. Through their ingenious works, all participants have showcased their innovation and professionalism, representing an emerging generation entrusted by the new era with the responsibility and mission of becoming designers, builders and bearers of ecological civilization.

We would like to extend our sincere gratitude to Delta Electronics for sponsoring International Solar Building Design Competition 2023.

This publication is dedicated to our fellow researchers and professionals who are devoted to advancing ecological civilization while promoting green, low-carbon development.

目 录
CONTENTS

阳光·零碳建筑　SUNSHINE & ZERO-CARBON ARCHITECTURE

2023台达杯国际太阳能建筑设计竞赛回顾
Review of International Solar Building Design Competition 2023

2023台达杯国际太阳能建筑设计竞赛评审专家介绍
Introduction to Jury Members of International Solar Building Design Competition 2023

获奖作品　Prize Awarded Works　　001

综合奖·一等奖·零碳设计项目
Comprehensive Awards · First Prize · Zero-Carbon Design Project

风·井园　Vertical Garden Shaped by Wind　　002

综合奖·一等奖·零碳提升项目
Comprehensive Awards · First Prize · Zero-Carbon Promotion Project

归零者——呼吸乐园　Zeroner-Breathing Paradise　　006

综合奖·二等奖·零碳设计项目
Comprehensive Awards · Second Prize · Zero-Carbon Design Project

低碳公园：生态绿谷　Low-Carbon Park: Ecological Green Valley	010
风·动　Wind Movement	014

综合奖·二等奖·零碳提升项目
Comprehensive Awards · Second Prize · Zero-Carbon Promotion Project

管巷风悦　Breeze Alley Oasis	018

综合奖·三等奖·零碳设计项目
Comprehensive Awards · Third Prize · Zero-Carbon Design Project

花间·绿意·新生　Flower, Green, Newborn	022
双子光盒　Gemini Light Box	026
风绿阡陌　The Windy Green Alley	030
风·巷·塔　Wind, Alley, Tower	034
乘风好去　Wind Catcher	038
光盒作用　Photosynthesis	042

综合奖・三等奖・零碳提升项目
Comprehensive Awards・Third Prize・Zero-Carbon Promotion Project

风光相伴，绿意同生	Breeze and Sunshine Accompany, Greenery and Architecture Coexist	046

综合奖・优秀奖・零碳设计项目
Comprehensive Awards・Honorable Mention・Zero-Carbon Design Project

零度科创馆・岭	Lingnan Zero Carbon Science and Technology Innovation Museum	050
拾级叠院	Clambered Up-Yard	054
绿光新谷	Green Valley	058
风之馆	Museum of Winds	062
接・融	Connection, Fusion	066
多级绿洲	Oasis	070
裂缝解码	GENO ART X	074
光之舞・荔园	Dance of Lights, Li Garden	078
转译竹筒屋	Bamboo Tube Transforming	082

管叠・风起 Ducts Play, Winds Arise	086
向阳而生 Growing towards the Sun	090
穿竹 Through the Bamboo	094
天上星河转 Turning Galaxy	098
风声不息 The Wind Never Dies Down	102

综合奖・优秀奖・零碳提升项目
Comprehensive Awards · Honorable Mention · Zero-Carbon Promotion Project

庭间序 Between Cavaedium	106
从游・木筑 Legolas Rotating	110
风廊・叶影 Wind Corridor, Leaf Shadow	114
还碳・环碳 Carbon Returning and Carbon Cycling	118
翕风纳绿 Delightful Wind Brings Greenery	122
风谷之竹 Bamboos in the Valley	126

综合奖·入围奖·零碳设计项目
Comprehensive Awards · Nomination · Zero-Carbon Design Project

天工开物	Buildings Torn Apart by Plants	130
风满楼	Wind Traveling	134
与风同巷	The Evolution of Cold Alley	138
入风吟	Chant in Wind	142
光檐风堂	Wind Hall under Solar Photovoltaics	146
星落	Meteor	150
生长·光谷	Growth, Optical Valley	154
风巷赏，光塔下思	Enjoying in Windy Alley, Thinking under the Tower of Light	158
羊城叠翠	Yangcheng Ecology	162
辉晖相映	Symphony of Solar and Snug	166
风之谷	The Valley of Wind	170
风巷·树影	Wind Alley, Trees Shadow	174
风正一帆悬	The Sail Hangs Straight Amid the Wind	178
创梦盒伙人	Dream Partners	182

无界方舟	The Unbounded Ark	186
风科谷	Technical Air Flue	190
山谷生长	Growing Valleys	194
梯山栈谷	Terraced Mountain Plank and Valley	198
光隐南园	Lingnan Garden Behind the Solar PV Panel	202
旋·浮	Spiral, Suspension	206
编影·竹光	Weaving Shadow, Bamboo Light	210
水境天梯	Waterfront Stairway	214
风穿绿廊 碳循未来	Wind Through Green Gallery, Carbon Recycling for the Future	218
风起·光旋	Wind Swirling with Light	222

综合奖·入围奖·零碳提升项目
Comprehensive Awards · Nomination · Zero-Carbon Promotion Project

叶园	The Vein	226
积木拼荫	Brick Puzzled Shadows	230
零碳方舟，"碳"未来	Zero-Carbon Ark, "Carbon" Surfing the Future	234

既改＋ All in One	238
竹升·凉岛 Rowing Bamboo, Cool Islands	242
光·帆 Light, Sail	246

有效作品参赛团队名单
Name List of All Participants Submitting Valid Works … 250

2023台达杯国际太阳能建筑设计竞赛办法
Guide for the International Solar Building
Design Competition 2023 … 260

2023台达杯国际太阳能建筑设计竞赛回顾
Review of International Solar Building Design Competition 2023

主题：阳光·零碳建筑

一、赛题设置

为积极响应"双碳"目标，推动校园建筑低碳化发展，不断提升建筑品质，2023台达杯国际太阳能建筑设计竞赛以"阳光·零碳建筑"为主题，设置零碳设计项目"广州科教城文化科技馆"和零碳提升项目"广州市公用事业技师学院社团综合楼"两个赛题，以推动可再生能源在建筑上的应用，探索零碳建筑的建设路径。

二、竞赛启动

2023年3月24日，2023台达杯国际太阳能建筑设计竞赛在北京启动。中国工程院院士、中国建设科技集团股份有限公司首席科学家、中国建筑设计研究院有限公司总建筑师、台达杯国际太阳能建筑设计竞赛评审专家组组长崔愷，中国建设科技集团股份有限公司副总裁、中国建设科技集团股份有限公司中央研究院执行院长刘志鸿，台达集团创办人暨荣誉董事长郑崇华，时任中国建筑设计研究院有限公司总经理马海，中国建筑设计研究院有限公司总工程师仲继寿，国家住宅与居住环境工程技术研究中心主任张磊等两百余人出席会议，共同开启本届竞赛。

Competition Theme: Sunshine & Zero-Carbon Architecture

I. Competition Tasks

International Solar Building Design Competition 2023 embraces the theme of "Sunshine & Zero-Carbon Architecture" to proactively address the objective of "Double Carbon" and enhance the low-carbon development of campus environment through continuous improvement in building quality. It establishes corresponding competition tasks as follows: Task 1: Cultural and Scientific-technological Museum of Guangzhou Science and Education Park, a project focusing on designing a new Zero-Carbon Architecture; and Task 2: The Complex Building for Students Societies of Guangzhou Public Utility Technician College, a project targeting the promotion of zero-carbon practices for the renovation of an existing property.

The two tasks aim to enhance the utilization of renewable energy in buildings, and explore a feasible approach to constructing Zero-Carbon Architectures.

组委会进行竞赛项目场地调研
The Organising Committee Conducts Research on Competition Venues

竞赛组委会专家与项目建设方进行沟通
Experts of the Competition Organising Committee Communicate with the Project Builder

崔恺院士（左三）、刘志鸿副总裁（右二）、马海总经理（右一）、仲继寿副总建筑师（左一）与郑崇华先生（左二）共同启动 2023 台达杯国际太阳能建筑设计竞赛
Academician Cui Kai (third from the left), Vice President Liu Zhihong (Second from the Right), General Manager Ma Hai (First from the Right), Associate Chief Architect Zhong Jishou (First from the Left) and Mr Zheng Chonghua (Second from the Left) launched the 2023 Delta Cup International Solar Energy Architecture Design Competition

II. Commencement of Competition

Competition 2023 was officially launched in Beijing on March 24, 2023. Over 200 professionals attended to witness the grand opening ceremony, including dedicated experts as follows:

Mr. Kai Cui, Academician of Chinese Academy of Engineering, Chief Scientist of China Construction Technology Group Co., Ltd. and Chief Architect of China Architecture Design and Research Group Co., Ltd. (hereinafter referred to as CADG), who also serves as the Jury Panel Leader of Competition 2023.

Mr. Zhihong Liu, Vice President of China Construction Technology Group Co., Ltd. (hereinafter referred to as CCTC), Executive Director of the Central R&D Institute of CCTC.

Mr. Chonghua Zheng, Founder and Honorary Chairman of Delta Electronics.

Mr. Hai Ma, General Manager of CADG.

Mr. Jishou Zhong, Chief Engineer of CADG.

Ms. Lei Zhang, Director of the National Engineering Research Center for Human Settlements of China.

竞赛海报（中、英文版）
Competition Poster (Chinese and English)

三、校园宣讲

自 2005 年第一届竞赛举办以来，组委会已先后前往清华大学、天津大学、东南大学、重庆大学、山东建筑大学等 60 多所高校开展竞赛校园宣讲活动，受到了高校师生的积极响应和好评。

5 月初至 7 月底，组委会前往东北大学、沈阳建筑大学、山东建筑大学、潍坊科技学院、重庆大学、广州大学、惠州学院、广东工业大学、兰州交通大学等高校举办了宣讲会并同步进行网络直播。组委会邀请可再生能源领域专家与师生们共同围绕太阳能建筑进行交流，并与师生分别针对在"双碳"目标下建筑行业发展及建筑相关专业毕业生工作方向等问题进行深入讨论，为师生们对建筑行业未来发展打开了新思路。系列宣讲会共吸引了千余名师生亲临现场参与，同时，近 6000 人次通过直播观看了宣讲内容，反响热烈。

同时，竞赛联合"迈向产能建筑"专家团队多次进行线上讲堂。内容涵盖了太阳能建筑技术应用现状与趋势、智能控制助力建筑低碳、天正建筑能耗及碳排放计算软件讲解、历届竞赛获奖作品分析和本届竞赛答疑等。通过线上讲堂，普及了低碳建筑技术，让观众们更深入了解竞赛赛题内涵，激发了参赛团队的设计灵感，对太阳能建筑应用技术进行了创新思考。通过线上讲堂，也提高了竞赛关注度和可再生能源领域相关知识积累，使竞赛宣讲会成为太阳能建筑领域行业知识、技术、理念的重要科普和交流平台。

III. Campus Roadshow

Since the inaugural of the Competition in 2005, the Organizing Committee has conducted campus roadshows at over 60 esteemed colleges and universities, including Tsinghua University, Tianjin University, Southeast University, Chongqing University, Shandong Jianzhu University, etc. These promotional activities have received positive responses from both faculty members and students.

From May to the end of July, the Organizing Committee conducted roadshow sessions along with on-line broadcasting in various academic institutions across China, including Northeastern University, Shenyang Jianzhu University, Shandong Jianzhu University, Weifang University of Science and Technology, Chongqing University, Guangzhou University, Huizhou University, Guangdong University of Technology, and Lanzhou Jiaotong University. The sessions successfully invited experts, educators and students from the renewable energy field to engage in in-depth discussions on the topic of solar-powered buildings. The educators and students of the universities and the Organizing Committee exchanged insightful views on the development of the construction industry and the career prospects for graduates with majors related to

与东北大学师生合影
Group Photo with Students and Faculty of Northeastern University

与沈阳建筑大学师生合影
Group Photo with Teachers and Students of Shenyang Jianzhu University

东北大学宣讲会现场
Lecture at Northeastern University

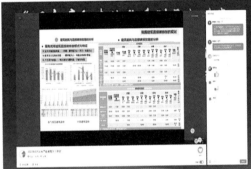
网络直播截图
Screenshot of the Webcast

与山东建筑大学师生合影
Group Photo with Teachers and Students of Shandong Jianzhu University

与重庆大学师生合影
Group Photo with Teachers and Students of Chongqing University

潍坊科技学院讲座现场
Weifang Institute of Science and Technology Lecture Scene

惠州学院讲座现场
Lecture at Huizhou College

与广州大学师生合影
Group Photo with Teachers and Students of Guangzhou University

重庆大学宣讲现场
Lecture at Chongqing University

与西安建筑科技大学师生合影
Group Photo with Teachers and Students of Xi'an University of Architecture and Technology

与广东工业大学师生合影
Group Photo with Teachers and Students of Guangdong University of Technology

线上宣讲会海报
Poster for Online Seminar

architecture, particularly in light of the potential influence brought forth by the national "dual-carbon" objective. These discussions generated fresh and valuable ideas for both faculty members and students regarding the future development of the construction industry. The sessions successfully attracted over a thousand educators and students to participate in on-site activities. Additionally, nearly 6,000 individuals joined through online broadcasting.

Concurrently, the Competition collaborated with an expert team from "Towards Productive Buildings" to conduct a series of online lectures encompassing various topics such as the current industrial landscape, trends in solar building technology application, intelligent control of low-carbon buildings, demonstrations of Tianzheng software for calculating building energy consumption and carbon emissions, analysis of winning works from previous competitions as well as Q & A sessions for Competition 2023. These online lectures effectively disseminated knowledge on low-carbon building technologies while providing participating teams with a comprehensive understanding of the contest tasks. Consequently, they facilitated

四、媒体宣传

自竞赛启动以来，组委会通过多种媒体形式开展竞赛宣传工作，包括：竞赛官方网站（双语）实时报道竞赛进展情况；线上线下同步开展太阳能建筑的科普宣传；在百度设置关键字搜索，方便大众查询，从而更快捷地登录竞赛网站；在新华网、腾讯网、新浪网、美联社、Archdaily 等50余家国内外网站上报道竞赛相关信息并链接到竞赛官方网站；与国内外多所校级媒体取得联系并发布竞赛信息与动态；通过微信公众号、微博等媒体平台实时发布竞赛相关动态，并提供竞赛相关资料下载与案例分析等，有效地提高了竞赛的影响力及参赛团队的技术能力。

the promotion and exchange of innovative thinking on the application of solar energy building technologies. The attention degree of the Competition has increased significantly through these lectures, while knowledge accumulation in the field of renewable energy and dissemination of relevant technical literacy have also been intensified. As a result, the Competition lectures have become an important platform for popularizing and exchanging industrial knowledge, technologies and ideas in solar building field.

IV. Media & Publicity

Since the launch of Competition 2023, the Organizing Committee has conducted publicity efforts through various media forms, including: real-time bilingual progress reports on the official website; simultaneous online and offline popularization of solar buildings ; keyword search in Baidu for easy public access to the official website; coverage on over 50 domestic and international websites such as Xinhuanet.com, Tencent, Sina.com, AP, Archdaily, etc., with direct links to the official website; liaisons with various domestic and foreign university-level media outlets for releasing competition information and updates;

Archdaily 国际版网站发布竞赛相关信息
Archdaily's International Website Publishes Information about Competitions

新华社发布竞赛相关信息
Xinhua News Agency Releases Information about the Competition

中国"学习强国"号发布竞赛相关信息
Information about the China Learning Power Number Release Competition

竞赛官网
Competition Website

《世界建筑》纸媒发布竞赛相关信息
Worldbuilders Paper Publishes Information on Competitions

real-time news through Wechat public account, Weibo and other media channels with downloadable information resources and case analysis. These endeavors effectively enhance both the influence of the Competition and technical capabilities of the participating teams.

V. Registration & Submission

The registration period of the Competition lasted from March 26, 2023 to August 15, 2023. A total of 733 teams successfully completed registrations through the official website, including 10 overseas teams from the United States, Canada, Singapore, Nigeria, etc. As of midnight (24:00) on September 15, 2023, the Organizing Committee has received a total of 249 submitted works, out of which 210 have been deemed valid.

VI. Entries Appraisal

The Entries Appraisal comprises four stages: formal examination, preliminary appraisal, middle appraisal and final appraisal.

From October 1st to 12th, the Organizing Committee arranged for experts to conduct a preliminary appraisal for valid works that passed the formal examination. Following the appraisal criteria outlined in the Competition Guide, the experts evaluated each work through compliance inspection, preliminary assessment, and scoring. A total of 100 works were selected for the middle appraisal procedure.

From October 13th to 27th, the Organizing Committee invited nine esteemed experts to assess the entries for middle appraisal. Through strict assessment by the experts, a selection of 62 works would be chosen and receive the Comprehensive Awards of the Competition; among them were 30 works that entered into the final appraisal procedure.

The Jury Panel for the final appraisal was composed of nine distinguished experts, including:

Mr. Kai Cui, Academician of Chinese Academy of Engineering; Mr. King

五、竞赛注册及提交情况

竞赛的注册时间为2023年3月26日至2023年8月15日，共733组团队通过竞赛官网完成注册，其中有来自美国、加拿大、新加坡、尼日利亚等境外注册团队10组。截至2023年9月15日24时，组委会共收到提交作品249件，其中有效作品210件。

六、竞赛评审

竞赛评审由形式审查、竞赛初步评审、中期评审及终期评审四个阶段组成。

10月1~12日，组委会组织专家对通过形式筛查的有效作品开展初评工作。根据竞赛办法中的评比标准，专家们对每件都作品进行了评审，经过合规性审查、竞赛初步评审及评分排名，共有100件作品进入中期评审。

10月13~27日，组委会组织9位知名专家对竞赛作品进行中期评审。经过专家组的严格评审，评分前62名的作品获得综合奖；其中，评分前30名的作品进入终期评审。

本届竞赛终期评审团队由中国工程院院士崔愷、马来西亚汉沙杨建筑师事务所创始人杨经文、澳大利亚技术与工程院院士Deo Prasad、荷兰代尔伏特理工大

Mun Yeang, Founder and President of T. R. Hamzah & Yeang Sdn. Bhd. of Malaysia; Mr. Deo Prasad, Academician of Australian Academy of Technological Sciences and Engineering; Mr. Peter Luscuere, Professor of Department of Architecture, Technology University Delf, Netherlands; Mr. Feng Qian, Professor of College of Architecture and Urban Planning Tongji University; Mr. Qiuping Huang, Chief Architect of East China Architectural Design & Research Institute Co., Ltd.; Mr. Ya Feng, Chief Consulting Engineer of China Southwest Architectural Design and Research Institute Co., Ltd.; Mr. Jishou Zhong, Chief Engineer of China Architectural Design and Research Co., Ltd.; Mr. Wenxing Jiang, Deputy General Manager, Building Automation Business Group, Delta Electronics Industry Co., Ltd.

16 works were selected from the 30 entries and ranked by the Jury Panel to enter the second stage of final appraisal: on-site evaluation session.

On December 4th, 2023, the evaluation session was held in Beijing in the form of on-site defense by students and panel discussions by the expert group. This format provided an excellent opportunity for students to effectively

终评专家组与答辩师生合影
Group Photo of the Final Assessment Panel with the Defending Teachers and Students

现场评审专家合影
Group Photo of on-site Evaluation Experts

学建筑系教授 Peter Luscuere、同济大学建筑与城市规划学院教授钱锋、华东建筑设计研究院有限公司总建筑师黄秋平、中国建筑西南设计研究院有限公司顾问总工程师冯雅、中国建筑设计研究有限公司总工程师仲继寿、台达电子工业股份有限公司楼宇自动化事业群副总经理江文兴 9 位专家组成。

经过终期评审专家组对前 30 名的作品评审打分，筛选出前 16 名的作品进入终期评审第二阶段——现场终期评审会。

2023 年 12 月 4 日，现场终期评审会在北京举行。会议采用学生现场答辩与专家组集中讨论的形式进行，旨在让学生更好地展示作品，也能帮助专家组更全面地了解与评价作品。上午，前 16 名的团队在评审现场通过图像、文字、语言、模型与视频等多元的方式展示、阐述作品，并回答专家组提问。专家组在评审时既注重作品创意，也考虑到作品中应用技术的适用性与可操作性。下午，国内外评审专家通过视频会议连线，历经多轮评选和讨论，最终选出竞赛的一等奖 2 项、二等奖 3 项、三等奖 7 项、优秀奖 20 项、入围奖 30 项，共计 62 项获奖作品。

showcase their works and allow the Jury Panel to evaluate them with a more comprehensive understanding. In the morning, the top 16 teams presented their works in succession using various methods such as images, text, oral expression, physical models and videos while addressing questions raised by experts. Throughout this process, both the novelty and applicability of the technologies employed in these works were taken into consideration by the Jury Panel. In the afternoon, domestic and foreign experts participated in panel discussion via video conference. After undergoing multiple rounds of evaluation and extensive discussions, a total of 62 exceptional works were successfully selected, comprising 2 first prizes, 3 second prizes, 7 third prizes, 20 honorable mentions, and 30 nominations.

现场评审专家组（左上）与马来西亚汉沙杨建筑师事务所创始人杨经文（右上）、荷兰代尔伏特理工大学教授 Peter Luscuere（下）等国际专家连线讨论
The On-site Judging Panel (top left) Discusses with International Experts, Including Yang Jingwen, Founder of Hansa Yang Architects, Malaysia (top right) and Peter Luscuere, Professor at Delft University of Technology, the Netherlands (bottom)

2023台达杯国际太阳能建筑设计竞赛评审专家介绍
Introduction to Jury Members of International Solar Building Design Competition 2023

评审专家
Jury Members

杨经文，马来西亚汉沙杨建筑师事务所创始人、**2016**梁思成建筑奖获得者
Mr. King Mun Yeang: Founder and President of T. R. Hamzah & Yeang Sdn. Bhd. of Malaysia; Winner of Liang Sicheng Architecture Award 2016

Deo Prasad，澳大利亚科技与工程院院士、澳大利亚勋章获得者、澳大利亚新南威尔士大学教授
Mr. Deo Prasad: Academician of Australian Academy of Technological Sciences and Engineering; Winner of the Order of Australia; Professor of University of New South Wales, Sydney, Australia

Peter Luscuere，荷兰代尔伏特理工大学建筑系教授
Mr.Peter Luscuere: Professor of Department of Architecture, Technology University Delf, Netherlands

崔愷，中国工程院院士、全国工程勘察设计大师、中国建筑设计研究院有限公司总建筑师
Mr. Kai Cui: Academician of Chinese Academy of Engineering; Master of National Engineering Survey and Design of China; Chief Architect of China Architecture Design and Research Group Co., Ltd. (CADG)

钱锋，全国工程勘察设计大师、同济大学建筑与城市规划学院教授、博士生导师

Mr. Feng Qian: Master of National Engineering Survey and Design of China; Professor and Doctoral Supervisor of College of Architecture and Urban Planning Tongji University (CAUP)

黄秋平，华东建筑设计研究有限公司总院总建筑师

Mr. Qiuping Huang: Chief Architect of East China Architectural Design & Research Institute Co., Ltd. (ECADI)

仲继寿，中国建筑设计研究院有限公司总工程师、中国建筑学会健康人居专业委员会和主动式建筑专业委员会主任委员

Mr. Jishou Zhong: Chief Engineer of China Architecture Design and Research Group Co., Ltd. (CADG); Chairman of Committee of Healthy Habita of the Architectural Society of China (ASC); Chairman of Committee of Active House (ASC) of the Architectural Society of China (ASC)

冯雅，中国建筑西南设计研究院有限公司顾问总工程师

Mr. Ya Feng: Chief Consulting Engineer of China Southwest Architectural Design and Research Institute Co., Ltd.

江文兴，台达电子工业股份有限公司楼宇自动化事业群副总经理

Mr. Roland Chiang: Deputy General Manager, Building Automation Business Group, Delta Electronics Industry Co., Ltd.

袁烽，同济大学建筑与城市规划学院教授、博士生导师、副院长

Mr. Feng Yuan: Professor, Doctoral Supervisor and Deputy Dean of College of Architecture and Urban Planning of Tongji University (CAPU)

彭晋卿，湖南大学土木工程学院教授、博士生导师、土木工程学院党委书记
Mr. Jinqing Peng: Professor, Doctoral Supervisor, and Secretary of the Party Committee of School of Civil Engineering, Hunan University

任军，天津大学建筑学院教授、天友建筑设计股份有限公司首席建筑师
Mr. Jun Ren: Professor of School of Architecture, Tianjin University; Chief Architect of Tenio Group

宋晔皓，清华大学建筑学院教授、博士生导师、副系主任，清华大学建筑学院建筑与技术研究所所长，清华大学建筑设计研究院副总建筑师
Mr. Yehao Song: Professor, Doctoral Supervisor and Deputy Dean of School of Architecture of Tsinghua University; Director of Architecture and Technology Institute of Tsinghua University; Deputy Chief Architect of Architectural Design and Research Institute of Tsinghua University (THAD)

张宏，东南大学建筑学院教授、博士生导师
Mr. Hong Zhang: Professor and Doctoral Supervisor of School of Architecture of Southeast University

刘恒，中国建筑设计研究院有限公司副总建筑师、绿色建筑设计研究院院长
Mr. Heng Liu: Deputy Chief Architect of China Architecture Design and Research Group Co., Ltd.(CADG); Director of Green Architecture Design and Research Institute of CADG

获奖作品

Prize Awarded Works

综合奖·一等奖·零碳设计项目
Comprehensive Awards · First Prize · Zero-Carbon Design Project

注册号：101860
Register Number：101860

项目名称：风·井园
Entry Title：Vertical Garden Shaped by Wind

作者：杜林涛、韩晨阳、毕雪皎、李洁、徐雪健
Authors：Lintao Du, Chenyang Han, Xuejiao Bi, Jie Li and Xuejian Xu

作者单位：天津大学
Authors from：Tianjin University

指导教师：严建伟、杨崴
Tutors：Jianwei Yan and Wei Yang

指导教师单位：天津大学
Tutors from：Tianjin University

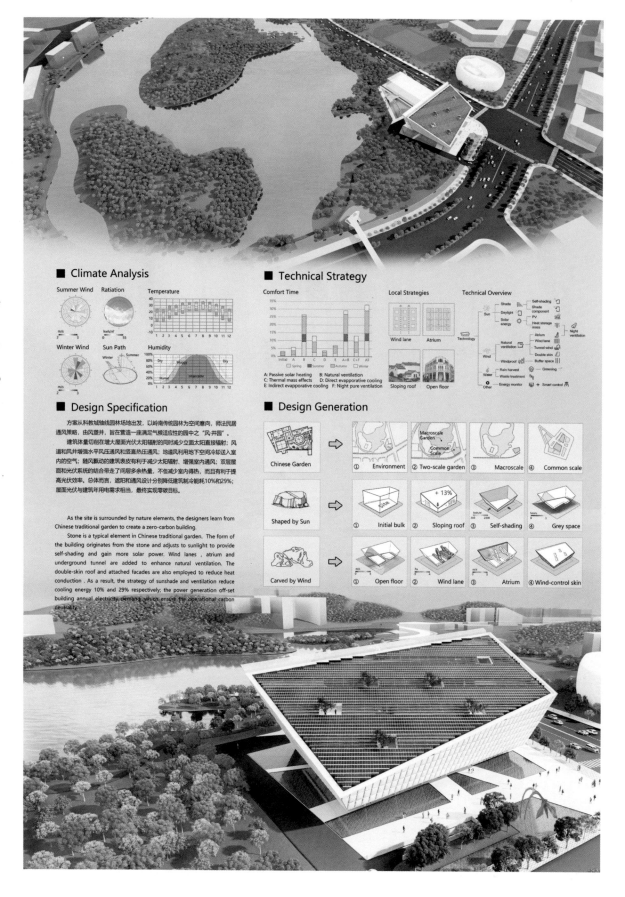

Function & Streamline

Exhibition

Public Space

Office / Research

Day 9:00-17:00
During the day time, both the north and south entrances are accessible, and people can choose to quickly reach each floor through the core tube or take the escalator to enjoy the view.

Night 17:00-24:00
After closing at night, the north atrium and roof garden can still be used independently, and the public can also walk in the wind lanes.

Site Plan 1:2000

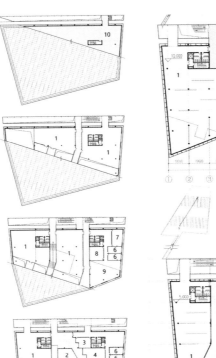

3rd Plan 1:500
1 Exhibition Hall
2 Science Activity
3 Office
4 Meeting Room

2nd Plan 1:500
1 Exhibition Hall
2 Office
3 Reception

4th-7th Plan 1:750
1 Exhibition Hall
2 Audiovisual Projection
3 Storage Room
4 Production Workshop
5 Maintenance Workshop
6 Design Research
7 Books and Materials
8 Meeting Room
9 Discussion Area
10 Roof Platform

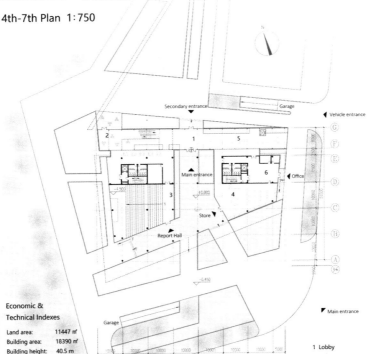

Economic & Technical Indexes
Land area: 11447 ㎡
Building area: 18390 ㎡
Building height: 40.5 m
Plot ratio: 0.6
Number of stops: 150

1st Plan 1:400
1 Lobby
2 Cafe
3 Report Hall
4 Store
5 Temporary Exhibition
6 Office Lobby

2023 台达杯国际太阳能建筑设计竞赛获奖作品集

专家点评：
该作品形体规整，向南侧倾斜的屋面巧妙增加了光伏的可用面积。面向湖的侧立面与环境和谐相融，高效的公共交通体系便捷实用。建筑中冷巷设计既富有趣味性，又注重自然通风与遮阳功能，有效地适应了广州气候，但南侧倾斜屋顶设计尚待提升，需更强调城市景观和谐性与场地道路的呼应。

Expert Commentary:
The work features a regular shape with a roof tilting towards the south side to cleverly increase the usable area for photovoltaic panel installations. The facade facing the lake seamlessly blends with the surrounding environment, while an efficient public transport system ensures its convenience and practicality. The incorporation of cold lanes into the design not only adds visual interest but also prioritizes natural ventilation and shading functionality, effectively adapting to Guangzhou's local climatic conditions. However, there is room for improvement in refining the design of sloping roofs on the south side, with greater emphasis needed on achieving harmony between urban landscape integration and responsiveness to adjacent roads.

Views of the Atrium

Views of Wind Lanes

■ Natural Ventilation

Wind Velocity

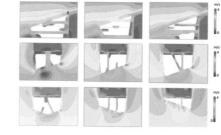

■ Climate Adaptability

Summer:
Vertical wind shaft

The high side window is opened and the low side window is closed to form the chimney effect.

Spring & Autumn:
Semi-outdoor space

All windows are open, and the space is integrated with the environment.

Winter:
Bufferzone

All windows are closed and combined with the heat storage mass to form a buffer zone.

The bottom of the PV is provided with an air chamber, which is open towards the summer dominant wind direction. And the air in the chamber is heated by the associated heat of the PV panel and discharged from the roof ridge. In addition the building can actively adjust the indoor wind and light environment by opening and closing the facade.

■ Note Details & Materials

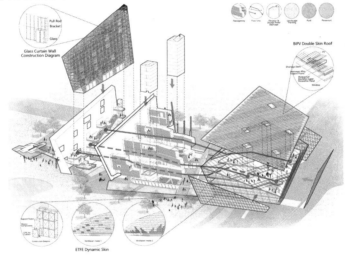

■ Energy & Carbon

Building Energy Consumption
Heating energy consumption is 1 kW·h/m²·a, cooling conumes 14 kW·h/m²·a, lighting consumes 15 kW·h/m²·a, other electric equipment consumes 18 kW·h/m²·a. Total energy consumption of this building is 49 kW·h/m²·a.

49 × 14000 = 686000 kW·h/a

PV Power Generation
The power efficiency of PV is 195 W/m². PVs laid at 20° produce electricity 241 kW·h/m²·a, while horizontally laid PVs produce electricity 231 kW·h/m²·a.

241 × 1900 + 231 × 1100 = 712000 kW·h/a

Operational Carbon Emission Balance
The carbon emission factor of electricity in Guangzhou is 0.8 $kgCO_2$eq/kW·h. This building can achieve zero carbon in operational stage.

548800 − 569600 = −208 $kgCO_2$eq/m²·a

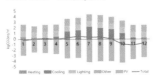

■ Effects of Strategies

Original

Step1: Self-shading
Effect: Cooling -10%

Step2: Ventilation
Effect: Cooling -29%

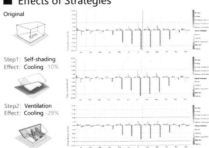

■ Indoor Temperature

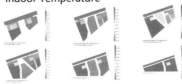

综合奖·一等奖·零碳提升项目
Comprehensive Awards · First Prize · Zero-Carbon Promotion Project

注册号：101903
Register Number：101903

项目名称：归零者——呼吸乐园
Entry Title：Zeroner—Breathing Paradise

作者：安莹、侯靖轩、胡维杭、张天岳、聂宇璇、张梦特
Authors：Ying An, Jingxuan Hou, Weihang Hu, Tianyue Zhang, Yuxuan Nie and Mengte Zhang

作者单位：天津大学
Authors from：Tianjin University

指导教师：朱丽
Tutor：Li Zhu

指导教师单位：天津大学
Tutor from：Tianjin University

Site Analysis

The transformation plan of the building achieves zero carbon through the combination of passive and active technology, and at the same time sets up stairs in the courtyard to connect the sky garden and the valley courtyard, extending from the bottom of the building all the way to the top of the building, forming a series of "vertical streets" connecting various functional areas, and deriving a number of different types of "places" with different functions in the middle.

Climatic Analysis

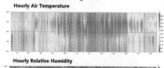

Analysis of People

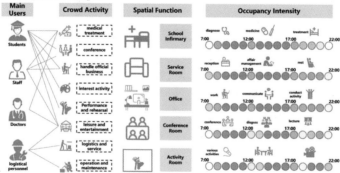
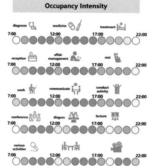

Concept of Technology

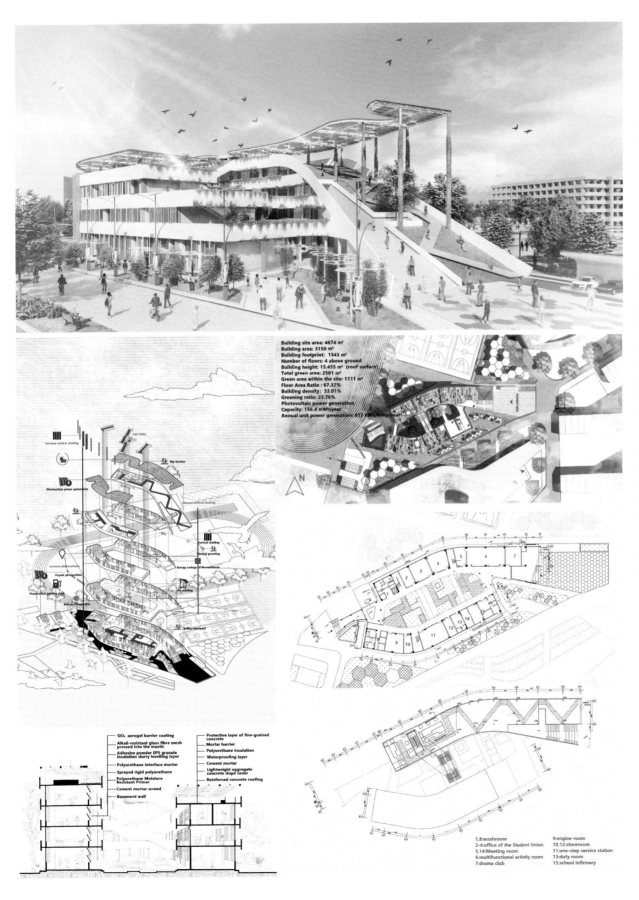

专家点评：

该作品将建筑设计和低碳、绿色进行了良好的结合，同时保持了建筑本体的完整性，利用垂直绿化实现建筑的被动降温。屋面光伏系统不干扰原有屋面的绿化和交通，并创造出连续、舒适的遮阴廊道。两个建筑之间的交通流线增强了形体之间的联系，增加了空间的趣味性，但建议对这一部分改造的可行性和经济性进行更深入的思考。

Expert Commentary:

The design successfully combines low-carbon principle and greenery into this renovation project, while preserving the building's original integrity. Passive cooling is achieved through the implementation of vertical greenery. The installation of a roof photovoltaic system does not compromise the existing landscape or hinder traffic flow on the roof; instead, it creates a continuous and comfortable shadowing corridor to strengthen these two factors. The connection and circulation between the two building blocks enhance their forms' cohesion while adding visual interest to space. However, further consideration should be given to the feasibility and cost-effectiveness in the renovation scheme, as recommended.

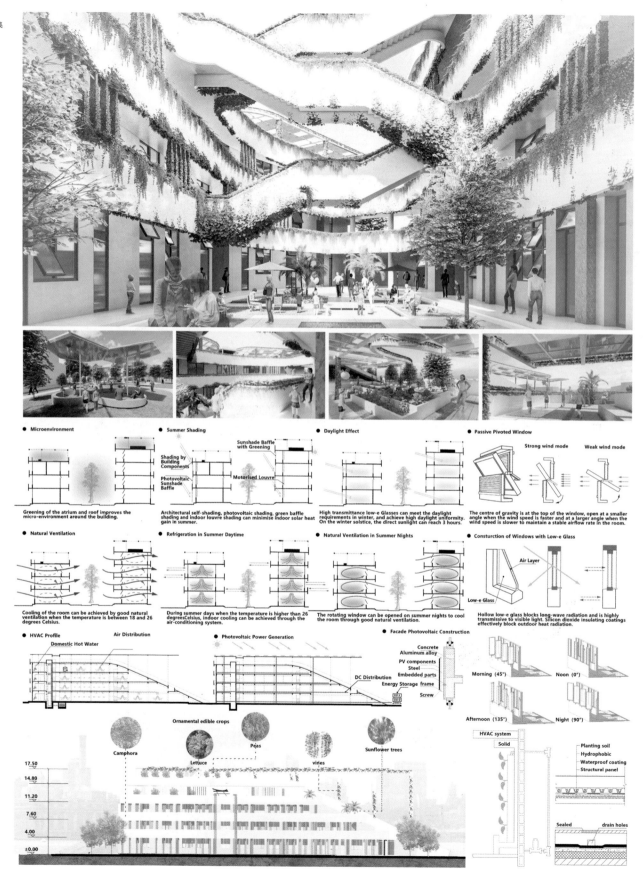

■ PVT+HVAC System

■ New HVAC System

■ PEDF System

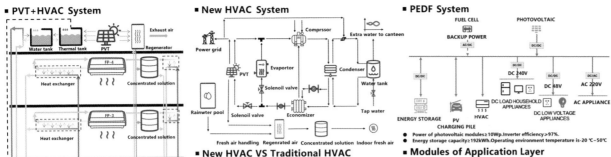

- Power of photovoltaic modules≥10Wp.Inverter efficiency≥97%.
- Energy storage capacity≥192kWh.Operating environment temperature is -20 ℃~50℃

■ New HVAC VS Traditional HVAC

■ Modules of Application Layer

- The data platform includes the management of energy consumption data and carbon emissions data, with a separate renewable energy operation monitoring system and a separate auxiliary decision-making system

■ Energy-efficient Operation

- Turn on the air conditioning unit one hour before the use of the building
- Before July 10: the chiller was set at 9℃; After August 30: the chiller was set at 9℃; July 10 to August 30: the chiller was set at 7 ℃

■ Energy Consumption Analysis

DESIGN BUILDING
ANNUAL TOTAL ENERGY CONSUMPTION

REFERENCE BUILDING
ANNUAL TOTAL ENERGY CONSUMPTION

ANNUAL HVAC ENERGY CONSUMPTION

ANNUAL HVAC ENERGY CONSUMPTION

MONTHLY ENERGY CONSUMPTION COMPOSITION

MONTHLY ENERGY CONSUMPTION COMPOSITION

■ Energy Consumption Comparison

Building waste heat recovery (kJ/day)		
	Summer	Winter
PVT heat utilization	895876.46	685082.00
Condensation heat recovery	6923664.00	4615920.00
Total	7819540.46	5301002.00
Solution Dehumidification Regeneration	1632960.00	544320.00
Domestic hot water supply	462274.56	1540915.20
Evaporator consumption	107505.18	82209.84
Total	2202739.74	2167445.04
Domestic hot water storage (canteen)	5616800.73	3133556.96

■ Energy Consumption Comparison

	ENERGY TYPE	HVAC	LIGHTING	ELECTRIC EQUIPMENTS	SUM TOTAL
DESIGN BUILDING	Subentry Energy Consumption (kW·h)	35582.4	55295.6	86184.5	177062.4
	Energy Consumption per Unit Area (kW·h/m²)	13.7	21.3	33.2	68.2
	ENERGY TYPE	HVAC	LIGHTING	ELECTRIC EQUIPMENTS	SUM TOTAL
REFERENCE BUILDING	Subentry Energy Consumption (kW·h)	56156.6	55295.6	86184.5	197636.7
	Energy Consumption per Unit Area (kW·h/m²)	21.6	21.3	33.2	76.1

Average energy consumption indicators for heating, cooling, and lighting in public buildings[kW·h/(m²·a)]					
Hot summer and warm winter areas	S_Office <20000m²	S_Office >20000m²	S_hotel <20000m²	S_hotel >20000m²	commercial building	School buildings
	34	58	95	94	148	31

- Compared with the reference building, the design building can save 10% of energy, and the HVAC energy consumption is reduced by 36%. The design of the air conditioning system and the material selection of the maintenance structure have remarkable energy-saving effects. The energy consumption of the building can meet the requirements of the code for the corresponding buildings in the corresponding area.

■ Daylight Autonomy

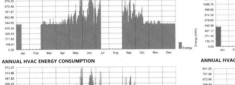

- Daylight autonomy (DA) results are shown in the figure. More than 60% of the area can ensure that the illumination is greater than 300 lx for more than 8h per day. In each room, the average lighting coefficient can reach a maximum of 17%, and the lighting coefficient of the other rooms also meets the standard provisions. The lighting uniformity of each room is between 0.10-0.27.

■ Carbon Emissions Calculations

Category	Carbon Emissions (TCO₂)
Operational Carbon Emissions	66.36
Renewable Energy Calculation	-58.62
Green House Gas Removal by Sinks	-11.87
Annual Carbon Emission	-4.13

Renewable Energy Calculation			
Solar panel area (m²)	Annual electricity production (kW·h)	Carbon emission factor (kgCO₂/kW·h)	Reduced carbon emissions (TCO₂)
968	156412	0.3748	58.62

Green House Gas Removal by Sinks			
Plant type	CO₂ Fixation Factors (kg/m²·a)	Area (m²)	CO₂ fixation (TCO₂)
Broad-leaved tree	22.5	322.9	7.26525
Deciduous tree	14.3	213.1	3.04733
Dioecious	8.15	109.8	0.89487
Ground cover plant	0.34	1939.3	0.659362
Sum total			11.866812

Operational Carbon Emissions			
Energy Type	Subentry Energy Consumption (kW·h)	Carbon emission factor (kgCO₂/kW·h)	Carbon emissions (TCO₂)
HVAC	35582.4	0.3748	13.34
Lighting	55295.6		20.72
Electric equipment	86184.5		32.3
Sum total	177062.4		66.36

■ Wind Environment of Typicalrooms

Office Plan / Integrated Service Room Plan / Activity Room Plan / Clinic Plan

Office Profile / Integrated Service Room Profile / Clinic Profile / Activity Room Profile

■ Baffle Opitimization

spacing 1m, angle 0° / spacing 0.7m, angle 0°
spacing 1m, angle 30° / spacing 0.7m, angle 30°
spacing 1m, angle 60° / spacing 0.7m, angle 60°
spacing 1m, angle 90° / spacing 0.7m, angle 90°

- The spacing and deflection Angle of the outdoor baffle were optimized, and the optimal result was 1m interval and 60° deflection. Interval 0.7m, deflection 90°. In the above two cases, the lighting uniformity can reach 23.2% and 22.4%, respectively, and the average annual maximum illumination is 2693lx.

■ Glass Material Comparison

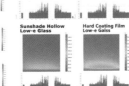

High Transmittance Hollow Low-e Glass / Double Silver Hollow Low-e Glass
Sunshade Hollow Low-e Glass / Hard Coating Film Low-e Galss

综合奖·二等奖·零碳设计项目
Comprehensive Awards · Second Prize · Zero-Carbon Design Project

注册号：101653
Register Number：101653

项目名称：低碳公园：生态绿谷
Entry Title：Low-Carbon Park：Ecological Green Valley

作者：张曦元、马圣新
Authors：Xiyuan Zhang and Shengxin Ma

作者单位：东南大学
Authors from：Southeast University

指导教师：寿焘
Tutor：Tao Shou

指导教师单位：东南大学
Tutor from：Southeast University

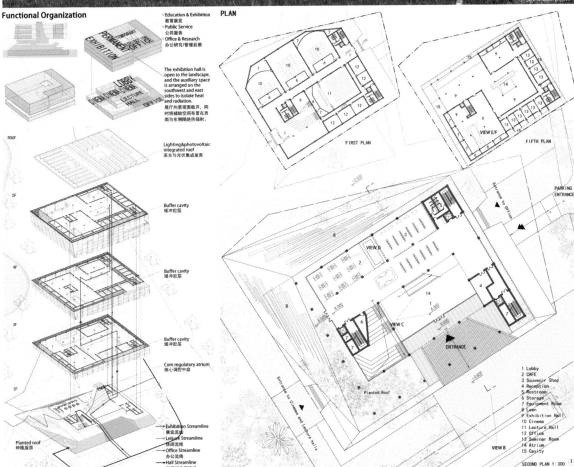

专家点评:

该方案形体与建筑空间的设计巧妙结合了自然地形、风环境、热环境与景观条件，回应场地的同时塑造出高品质的公共空间。通过空间组织与环境调控的整合，将环境调控区与气候缓冲区相结合，满足建筑内部不同空间对功能和环境性能的要求。通风采光一体化腔层以及其他通高空间解决了场馆内部通风与采光的问题，展现了设计的整体性。

Expert Commentary:

The architectural form and spatial design of the scheme have been meticulously crafted, taking into account crucial factors such as natural terrains, wind patterns, thermal environment and landscape characteristics. This results in a high-quality public space that effectively responds to the site conditions. Through integrating spatial organization and environmental regulation measures, the environmental regulation zone and climate buffer zone are successfully combined to meet both functional requirements and environmental performance standards across various areas within the building. In addition, by incorporating the integrated cavity layers along with high-ceiling spaces, the venue's ventilation efficiency is enhanced while optimizing natural lighting, showcasing an integral thinking behind the design.

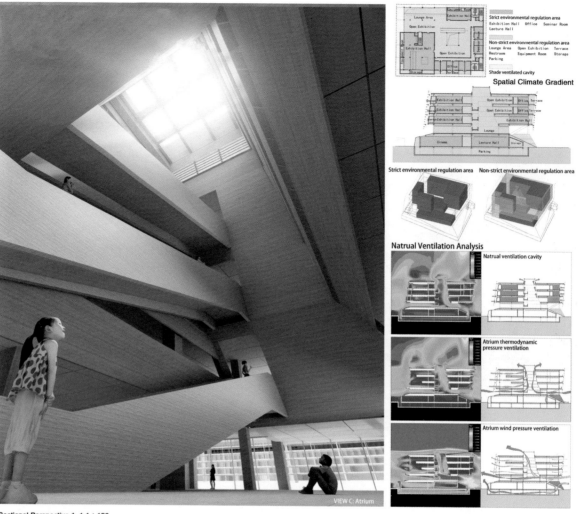

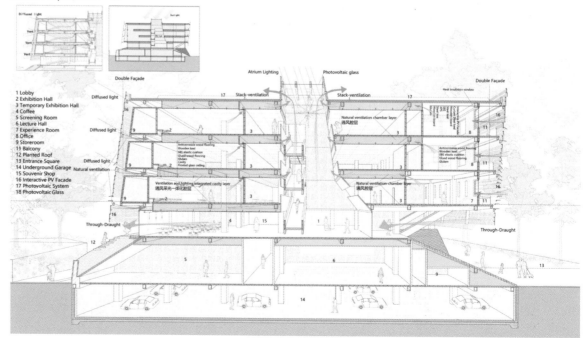

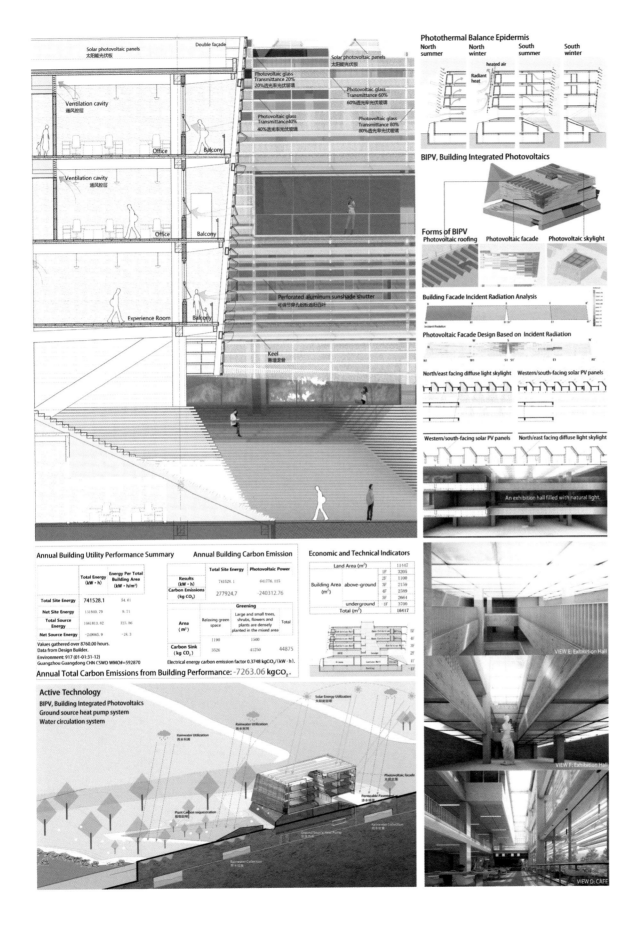

综合奖·二等奖·零碳设计项目
Comprehensive Awards · Second Prize · Zero-Carbon Design Project

注册号：101761
Register Number：101761

项目名称：风·动
Entry Title：Wind Movement

作者：冯晓潼、靳立越、张珏澜、陈安娴
Authors：Xiaotong Feng, Liyue Jin, Yulan Zhang and Anxian Chen

作者单位：重庆大学
Authors from：Chongqing University

指导教师：何宝杰
Tutor：Baojie He

指导教师单位：重庆大学
Tutor from：Chongqing University

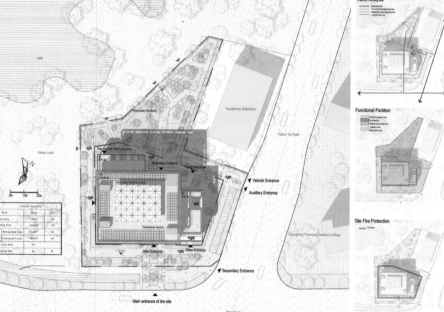

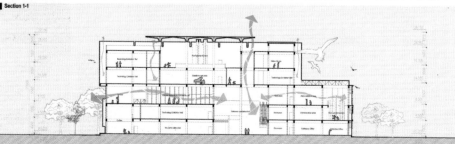

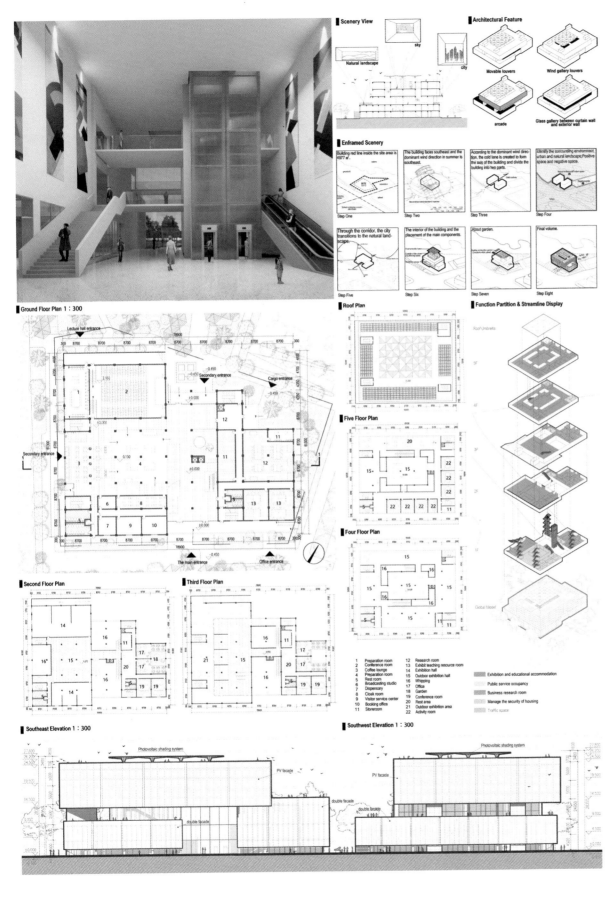

专家点评：

该方案设计手法成熟，空间完整，对场地环境和肌理的回应到位。对冷巷、骑楼、满洲窗、天井等本土元素的转译效果较好，综合考虑了材料选择、光伏应用、遮阳隔热等，技术运用整体性很强。但立面封闭性较强，内部需要空气调节的空间过多，自然通风的影响范围有待扩大。

Expert Commentary:

The scheme presents refined design skills while offering a complete spatial arrangement and a thoughtful response to the site environment and architectural fabric. Local traditional elements, such as cold lanes, overhanging corridors, Manchurian windows and patios are intelligently transformed and integrated into its architecture expression of the scheme. Additionally, the scheme includes comprehensive considerations encompassing material selection, photovoltaic application, shading treatments and heat insulation. This creates a robust integral feature in terms of technological applications. However, it is worth noting that the facades are overly enclosed which increases reliance on interior air conditioning systems for too many spaces. Therefore, it is recommended to enhance the incorporation of natural ventilation by reducing the areas requiring air conditioning in the design.

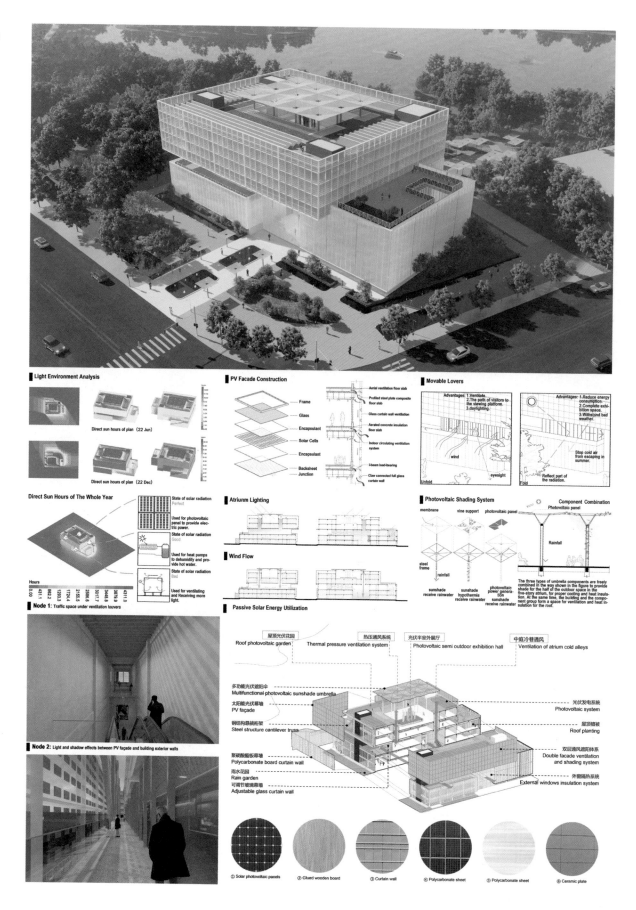

综合奖·二等奖·零碳提升项目
Comprehensive Awards · Second Prize · Zero-Carbon Promotion Project

注册号：101753
Register Number：1101753

项目名称：管巷风悦
Entry Title：Breeze Alley Oasis

作者：吴奕莹、贺川、颜廷旭
Authors：Yiying Wu, Chuan He and Tingxu Yan

作者单位：多伦多大学、Farrow Partners、Stepin 设计工作室
Authors from：University of Toronto, Farrow Partners and Stepin Desig Studio

管巷风悦
Breeze Alley Oasis

学生社团楼节能改造受限于建筑形态，难以应用被动节能手段。为适应广州气候，以岭南建筑为借鉴，我们提出将建筑与庭院结合，注重自然通风、遮阳隔热、绿植调节。打开楼梯间形成冷巷，结合下向式通风改善中庭热舒适度，通过热压送入室内。

光伏板是实现零碳的必要手段，但受限于现有覆土屋顶。通过遗传算法平衡光伏板面积、采光、遮阳三个相互制约的目标生成新屋顶。形体上，保持建筑外立面与校园风格统一，削弱屋顶体量与庭院内立面呼应，营造轻松的氛围。

The energy-efficient renovation of the student association building is constrained by its architectural form, making it difficult to apply passive energy-saving measures. To adapt to the climate of Guangzhou and draw inspiration from Lingnan architecture, we propose integrating the building with courtyards, emphasizing natural ventilation, sun shading, and greenery regulation. By opening stairwells to create cold corridors and incorporating downward ventilation, we aim to improve thermal comfort in the atrium and introduce heat from the atrium into the interior.

Photovoltaic panels are essential for achieving carbon neutrality but are limited by the existing green roof. Using genetic algorithms, we balance the area of photovoltaic panels, daylighting, and shading to generate a new roof design. Aesthetically, we maintain the architectural facade's consistency with the campus while reducing the roof's volume to harmonize with the courtyard's interior facades, creating a relaxed atmosphere.

Site Analysis

场地现状：建筑本体遮挡通风、北侧大面积操场无遮阴，需实现零运行碳排放。

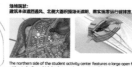

The northern side of the student activity center features a large open field without shading, which leads to an increase in ambient air temperature around the building during the summer. During preliminary wind simulations, the site was laid out in an east-west direction, resulting in the prevailing south-to-north winds being blocked by the building itself. Which creates poor turbulent airflow circulation within the courtyard. The building need to enhance its energy efficiency to achieve zero operational carbon emissions.

Net-Zero Design Process

受现有建筑限制，需改善微气候，增建屋顶以铺设光伏。

In the process of designing a zero-carbon building, there's a need to reduce energy consumption and obtain energy from renewable sources. Student commons face challenges due to the already constructed building, including an open atrium that hampers heat recovery ventilation and building orientation hindering natural ventilation.

The first step involves changing the microenvironment of the building site. Secondly, improving the building envelope's energy efficiency is crucial. Efficient HVAC and monitoring systems can help reduce energy demand. Another challenge is integrating photovoltaic panels, which may require additional roof construction to meet energy needs.

First Floor Plan

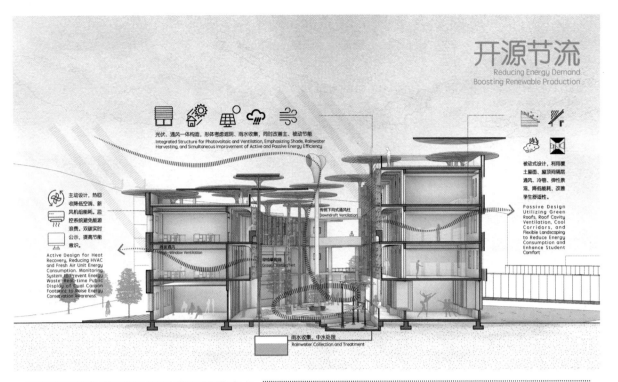

专家点评：

该方案巧妙地结合建筑改造与技术应用，独立系统设计不影响既有建筑施工。建筑外观轻巧活泼，适应校园社团综合楼主题，优化采光通风。采用冷巷策略创新解决微气候问题，通过遗传算法平衡采光、遮阴、光伏面积的目标思路新颖独特，风管引进自然风设计有趣实用。

Expert Commentary:

The scheme ingeniously incorporates technology applications in the building renovation by utilizing an independent system that minimizes potential impact on the ongoing construction of the existing property. The exterior of the building features a gentle and lively aesthetic, aligning with the theme of a campus community complex while optimizing lighting and ventilation effects. The innovative implementation of a cold lane strategy effectively addresses the micro-climate concerns. A genetic algorithm is employed to achieve a balanced objective that encompasses lighting, shading and photovoltaic area calculation, presenting a novel approach. The design of air ducts for introducing natural wind is both captivating and practical.

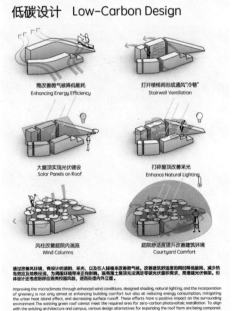

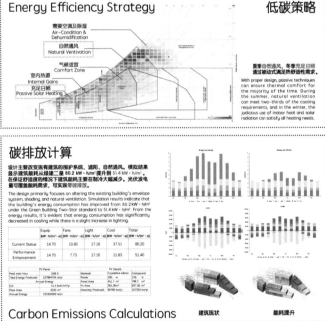

被动降碳
Passive Techniques for Carbon Emission Reduction

Step 1. 打开楼梯间引入新风 Open the stairwell for introducing fresh air

冷巷设置在现有三处楼梯间，通过活动墙板，开关冷巷

夏季打开墙体，使得楼梯间作为冷巷为中庭提供自然风，同时竖向遮阳板还能够遮挡西晒。冬季湿冷，板壁关闭，减少冷空气流通。

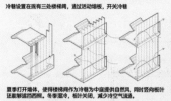 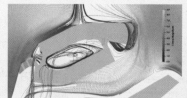 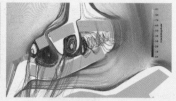

The building's wind environment has noticeably improved, as airflow passes through the cooling alley formed by the stairwell and enters the atrium. This helps enhance ventilation in the south side of the building and the of the site.

Step 2. 改善中庭湍流 Improving Atrium Turbulence

利用传统捕风塔下向式通风技术

However, wind simulation results showed turbulence within the atrium. To address this, we employed downward ventilation techniques to direct rooftop airflow to the level of human activity.

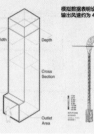 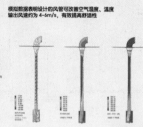

模拟数据表明设计的风管可改善空气湿度、温度，输出风速约为4-6m/s，有效提高舒适性

Step 3. 引风入室 Introduce Air Indoors

气候炎热时，关闭外侧窗户，打开高窗和中庭窗户，中庭绿植、通风设计可降低温度。新鲜空气从窗户进入，热压将热气从高窗排出，避免接纳太阳光无遮阴晒场地加热的空气进入室内。

气候寒冷时，关闭窗户和高窗，空斗墙有助保温。气候适宜时打开两层窗户对流通风。

During hot weather, the outer windows are closed, and the high windows and atrium windows are opened. The presence of greenery and ventilation design in the atrium helps lower the temperature. Fresh air enters through the windows, while the Venturi effect expels heated air from the sun-exposed open areas of the courtyard from entering indoors.

光伏柱、风管构造
Photovoltaic Column and Ventilation Duct Structures

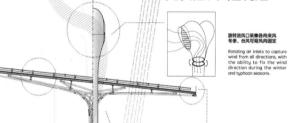

旋转进风口采集各向来风，冬季、台风可背风向固定
Rotating air inlets to capture wind from all directions, with the ability to fix the wind direction during the winter and typhoon seasons.

屋顶倾角5°冲刷泥污，雨水定向引流收集
The roof slope of 5 degrees facilitates the runoff of dirt and debris, while directing rainwater for collection.

由铝型材承托光伏板固定在钢柱上，可从底部检修、更换
Photovoltaic panels are supported by aluminum profiles and secured on steel columns, allowing for maintenance and replacement from the underside.

柱头加箍固定在现有建筑柱子上
Reinforced column heads are fixed onto existing building columns.

钢柱深埋入地下，应对水平受力
Steel columns are deeply embedded underground to withstand horizontal forces.

主动降碳
Active Techniques for Carbon Emission Reduction

Roof Design / Sunshade & Daylighting

通过遗传算法，平衡采光、遮阳、光伏面积三个矛盾的目标，程序生成上千个结果以与现有建筑协调为标准，人为选出最优解。

遮阳 采光 光伏面积
Shading / Daylighting / PV Panel Area

Using genetic algorithms to balance the conflicting objectives of daylighting, shading, and photovoltaic area. After thousands of simulations, the computer balances the target requirements and selects some superior solutions. The criteria for selecting the optimal solution are based on harmonizing with school.

传统岭南建筑特点 / Building Envelope

强调自然通风、遮阳隔热，形成与庭院结合的开敞空间。
墙体构造参考传统岭南建筑空斗墙

Traditional Lingnan architecture in Guangzhou places a strong emphasis on natural ventilation, sun shading, and the creation of open spaces integrated with courtyards. This approach is designed to enhance comfort and sustainability by working in harmony with the local climate.

墙中保持空气流通能够减少水汽向内渗入，并且空气或有很好的隔热作用。遮阳层的穿孔金属板能够彻彻底底的反射日照光，减少墙面对阳光的热量吸收。

光伏系统满足零碳需求 / Photovoltaic Power Generation

经计算光伏发电量可以满足零碳需求。光伏屋顶由风柱和智能钢柱两种方式承托，通过屋顶或是屋面与现有建筑结合。

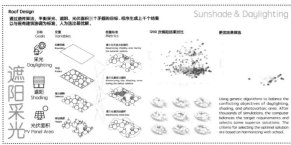

建筑适性改善
增加光伏板后建筑采光受影响较小，但添加后仍满足室内用光需求并光大幅改善。建筑使用者对于减少太阳直射和拉窗帘增加室内照明。

After adding photovoltaic panels, there is a slight impact on the building's daylighting, but it still meets indoor lighting requirements. Glare is significantly reduced, preventing occupants from needing to close curtains to reduce direct sunlight and increase interior lighting.

极端暴雨温区，采光要求同时设置透光光伏板与不透光光伏板。化合物型为广州阴雨天气，利用色光发电同时，避免部分遮挡带来安全隐患。无镉吸收体选品且电池器高发电效率。

建筑产能

The photovoltaic roof is supported in two ways: by wind columns and conventional steel columns. These structures are fixed to the roof or integrated with the existing building, considering shading and lighting requirements. Both transparent and opaque photovoltaic panels are used.

围护结构材料固碳量分析

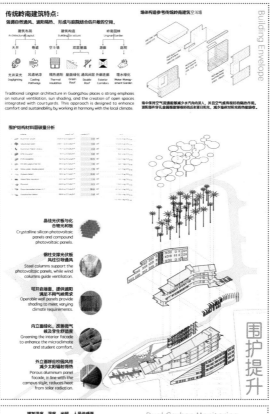

- **晶硅光伏板与化合物光伏板**
  Crystalline silicon photovoltaic panels and compound photovoltaic panels.

- **钢柱支撑光伏板风柱引导通风**
  Steel columns support the photovoltaic panels, while wind columns guide ventilation.

- **可开启墙面，提供遮阳满足不同气候需求**
  Operable wall panels provide shading to meet varying climate requirements.

- **内立面绿化，改善微气候及学生舒适度**
  Greening the interior facade to enhance the microclimate and student comfort.

- **外立面呼应校园风格减少太阳辐射得热**
  Porous aluminum panel facade, in line with the campus style, reduces heat from solar radiation.

围护提升

Sponge City / 海绵城市

通过增加绿地、改善微气候，构造一个有生机的、可持续的弹性景观在控制内管造一个新的自然环境。

The landscape design for this project involves the transformation of rigid paving, the expansion of green areas, and the creation of a pleasant microclimate, all while providing a series of public spaces conducive to social activities.

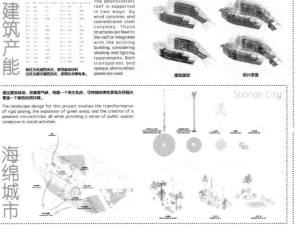

Dual Carbon Monitoring & Control System / 智能监控

- 增加温度、湿度、光照、人员传感器异常报警，并与设备联动减少能耗。
  Adding temperature, humidity, light, and occupancy sensors for anomaly detection and energy consumption reduction through equipment integration.

- 自动控制系统，预设设备使用场景，增加照明插座。
  An automated control system with preset usage scenarios, additional lighting circuits, and zoning controls for air conditioning and lighting to avoid unnecessary energy consumption.

- 能源统计分析，运行规程中排放数据展示、光伏绿电发电数据展示。
  Energy statistics analysis, displaying carbon emissions data in operation guidelines, as well as photovoltaic green energy generation data to promote energy-conscious behavior among students.

综合奖·三等奖·零碳设计项目
Comprehensive Awards · Third Prize · Zero-Carbon Design Project

注册号：101330
Register Number：101330

项目名称：花间·绿意·新生
Entry Title: Flower, Green, Newborn

作者：牛嘉琪、张燕燕、卢韵莹、倪英杰
Authors：Jiaqi Niu, Yanyan Zhang, Yunying Lu and Yingjie Ni

作者单位：中国矿业大学
Authors from：China University of Mining and Technology

指导教师：段忠诚、马全明、邵泽彪
Tutors：Zhongcheng Duan, Quanming Ma and Zebiao Shao

指导教授单位：中国矿业大学
Tutors from：China University of Mining and Technology

■ Backround Introduction

■ Site Conditions

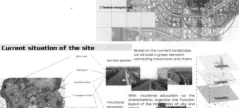

■ Annual Wind Rose ■ Wind Rose of 12 Months

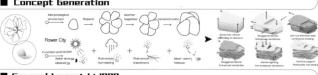

③ Psychrometric Chart

■ Concept Generation

■ General Layout 1:1000

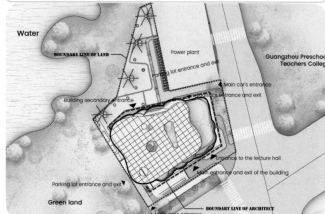

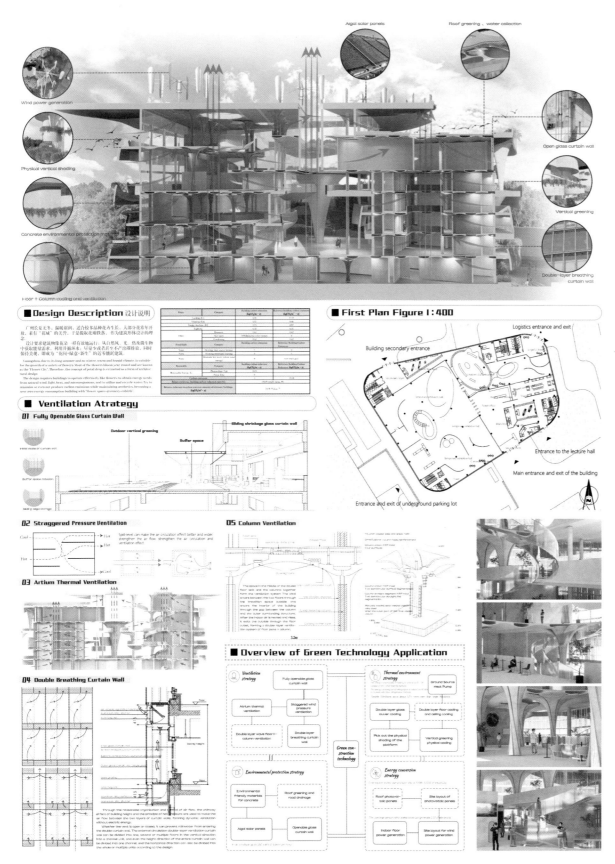

专家点评：

该方案形态设计契合科技馆主题，曲线与圆角元素增添了空间趣味。立面融合遮阳设计，外观丰富统一。波浪楼板连接柱子的通风设计新颖，强化了自然通风效果，带走日间热量，营造了舒适热环境。整体设计富有创意，空间利用高效。

Expert Commentary:

The scheme's form aligns with the theme of a science and technology museum, incorporating curves and rounded corner elements to add spatial dimensions and visual interests. The facade seamlessly integrates shading design, creating a diversified exterior appearance in a harmonious way. The innovative inclusion of wave floors connecting column ventilation enhances natural airflow, effectively dissipating daytime heat and creating a comfortable thermal environment. Overall, the design showcases creativity and optimizes space efficiency.

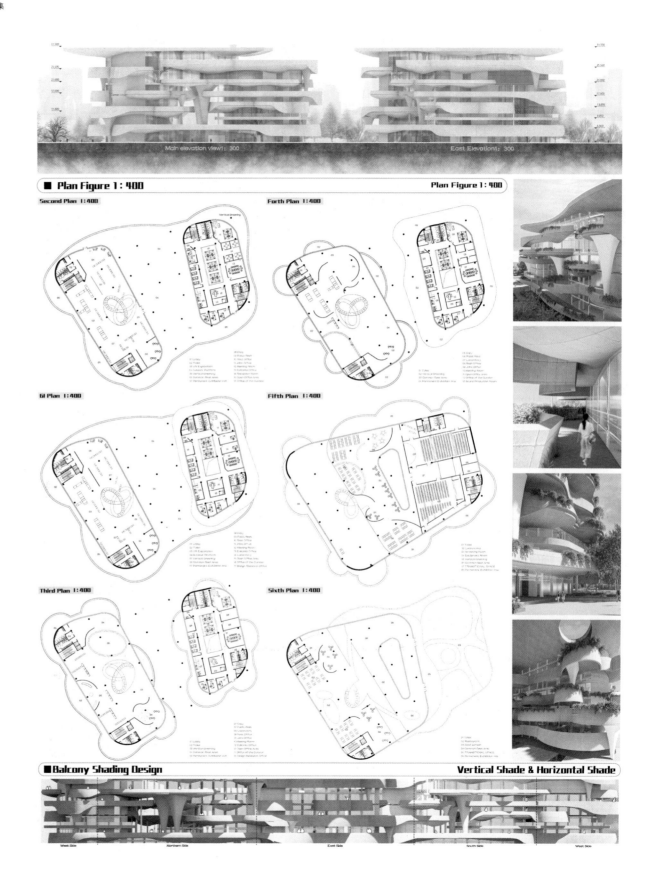

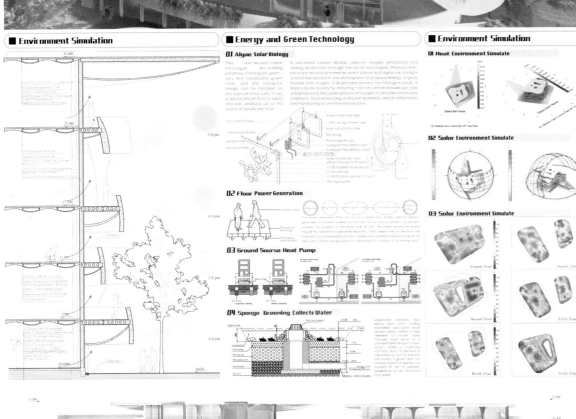

综合奖·三等奖·零碳设计项目
Comprehensive Awards · Third Prize · Zero-Carbon Design Project

注册号：101373
Register Number：101373

项目名称：双子光盒
Entry Title：Gemini Light Box

作者：陈嘉贝、赵玥、李啸跃、李昱瑶
Authors：Jiabei Chen, Yue Zhao, Xiaoyue Li and Yuyao Li

作者单位：山东建筑大学
Authors from：Shandong Jianzhu University

指导教师：侯世荣
Tutor：Shirong Hou

指导教师单位：山东建筑大学
Tutor from：Shandong Jianzhu University

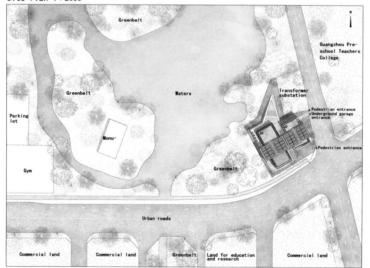

基地位于广州市增城区，南亚热带海洋性季风气候，气温高、雨量充沛、霜日少、光照充足。因此，方案集中在遮阳与太阳能利用、室内外风环境方面。
在形体上，本次设计充分利用周边优质景观面。除大面积室外观景坡道和休憩平台外，顶部的折形光伏发电一体式的遮阳大屋顶既利用文丘里效应改善通风环境，又美化了整个建筑形体。
在功能上，考虑到周边商业用地、教育科研用地和学校用地等多种属性，本次设计将科技馆定义为"面向多主体、呼应各环境"的"研、展、育一体式"场所，旨在便利城市各年龄段人群的科普教育学习，形成片区文化效应。
此外，建筑双表皮（穿孔板和可调节格栅）、屋顶绿化、蓄水屋面、遮阳光伏发电的一体式大屋顶使得建筑中的绿色低碳技术可视化，在科技普及的同时实现低碳技术的可展示性。

The base is located in Zengcheng District, Guangzhou, with a tropical maritime monsoon climate in South Asia, with high temperature, abundant rainfall, few frost days and sufficient sunshine. Therefore, the scheme focuses on shading and solar energy utilization, indoor and outdoor wind environment.
In terms of shape, this design makes full use of the surrounding high-quality landscape surface. In addition to the large-scale outdoor scenic ramp and rest platform, the large shading roof with integrated photovoltaic power generation at the top not only uses the venturi effect to improve the ventilation environment, but also beautifies the entire building shape.
In terms of function, taking into account the multiple attributes of surrounding commercial land, education and research land and school land, this design defines the science and technology museum as a "research, exhibition and education" place that is "oriented to multiple subjects and responds to various environments", aiming to facilitate the popular science education and learning of people of all ages in the city and form a cultural effect of the area.
In addition, the integrated large roof with double skin (perforated panels and adjustable grille), roof greening, water storage roof, and shading sunlight power generation visualizes the green and low-carbon technology in the building, and realizes the showability of low-carbon technology while popularizing technology.

Dry Bulb Temperature

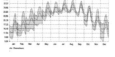
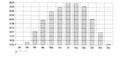
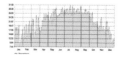

Scenario Generation

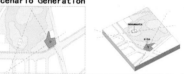
Step1. Determine the axis order of the building as a whole according to the existing monuments, green spaces, water areas and building red lines.

Step2. Based on the axis, science and technology exhibitions and office scientific research blocks are set up, leaving gaps to achieve visual penetration.

Step3. A transverse glass block is placed to connect the two blocks, protruding outward to echo the environment on both sides of the site.

Step4. Set up a viewing and leisure platform, make full use of the landscape of water and green space, and form a good landscape viewing surface.

Step5. Due to the environmental and climatic characteristics of the site, a sun-shading integrated roof is set to make full use of the Venturi effect.

Step6. Improve the site design, combine the surrounding waters to form water and land winds to improve the microclimate around the building.

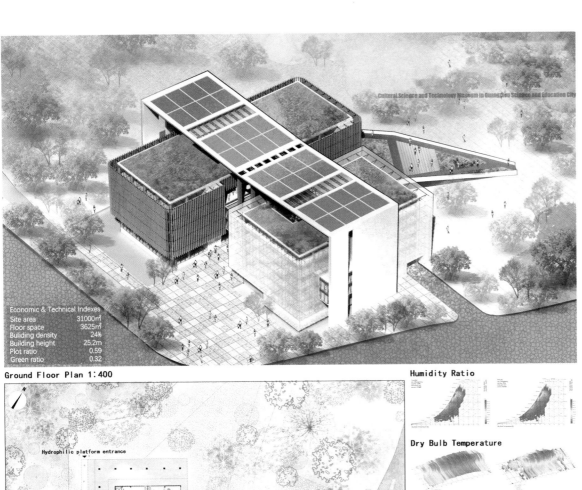

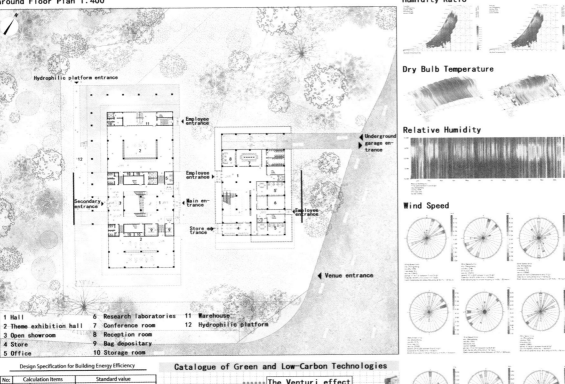

专家点评：
该方案建筑设计手法得当，将建筑划分成两个主要体块并巧妙形成通廊，在北侧设置的观景平台对场地与周边环境进行了适宜的回应。遮阳与热压通风设计适应气候特点，但外层折叠屋顶结构表达不足，形体略显单薄。建议加强结构设计，提升整体稳定性。

Expert Commentary:
The architectural design of the scheme showcases appropriate skills, effectively dividing the building into two blocks to create corridor spaces. The observation platform on the north side aptly responds to the site and its surrounding environment. The design incorporates shading and thermal natural ventilation to adapt to local climatic conditions. However, the structure of the folded roof is not sufficiently expressed, while the overall shape spears slightly thin. It is also recommended to reinforce the structural design to enhance general stability.

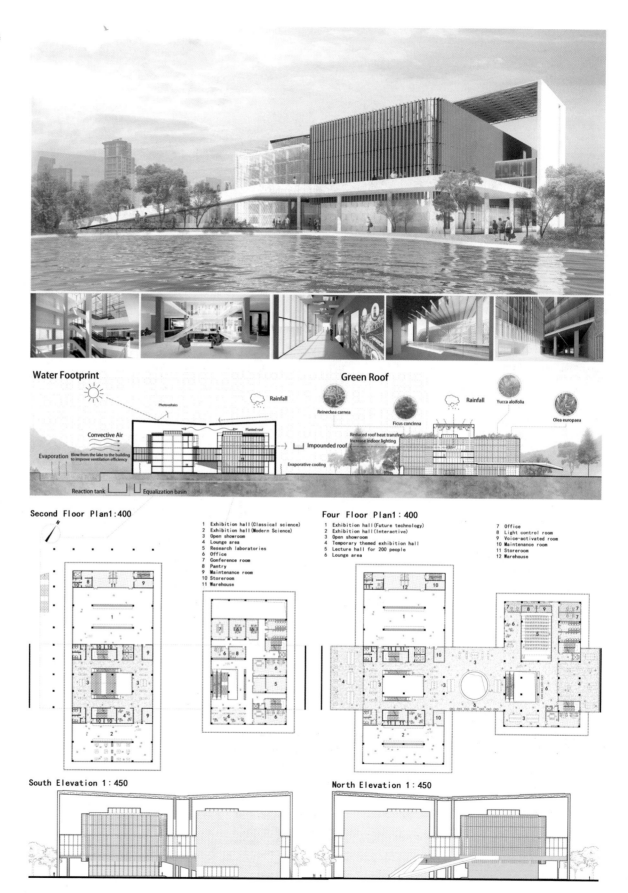

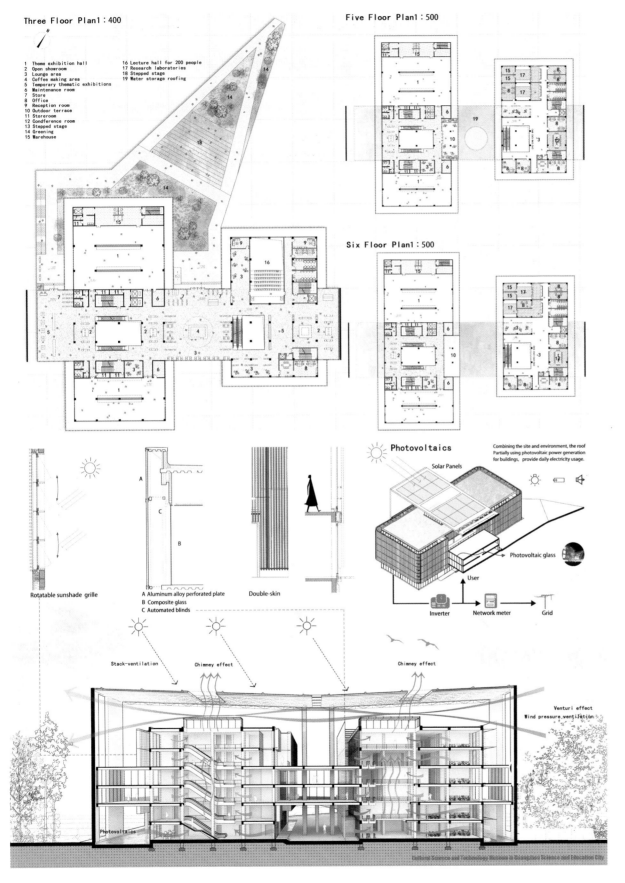

综合奖·三等奖·零碳设计项目
Comprehensive Awards · Third Prize · Zero-Carbon Design Project

注册号：101425
Register Number：101425

项目名称：风绿阡陌
Entry Title：The Windy Green Alley

作者：徐艳芳、李莹、方溢凯
Authors：Yanfang Xu, Ying Li and Yikai Fang

作者单位：厦门大学
Authors from：Xiamen University

指导教师：贾令堃、石峰
Tutors：Lingkun Jia and Feng Shi

指导教师单位：厦门大学
Tutors from：Xiamen University

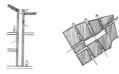

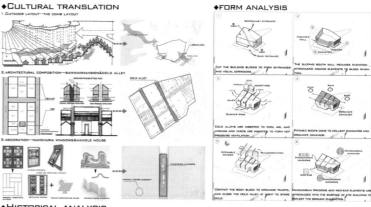

◆ DESIGN EXPLANATION

设计旨在回应场地气候、环境以及文化，利用梳式建筑布局手法以及冷巷原理，并结合架空形成视觉通廊，营造丰富、绿色的建筑空间。在零碳策略上，力求建筑全生命周期低排放，运行阶段分为节能、产能、提高能源利用率、增加碳汇以及可调监控策略，并选择海藻砌块、当地废弃甘蔗秆再造木材等环保材料，实现零碳。设计中采取被动、主动优化的方式，从建筑形体、表皮、构造等方面进行设计和应对，充分解决环境热湿、通风问题。

The design is designed to respond to the climate, environment and culture of the site, using the comb building layout and the principle of cold lane, and combining with the overhead to form a visual corridor to create a rich green architectural space. In terms of zero-carbon strategy, we strive for low emissions in the whole life cycle of the building. The operation stage is divided into energy saving, production capacity, energy efficiency improvement, carbon sink increase and adjustable monitoring strategy, and environmental protection materials such as seaweed blocks and local waste sugarcane rods are selected to achieve zero carbon. In the design, passive and active optimization methods are adopted, and the design and response are carried out from the aspects of building form, skin and structure, so as to fully solve the problems of environmental heat and humidity and ventilation.

◆ HISTORICAL ANALYSIS

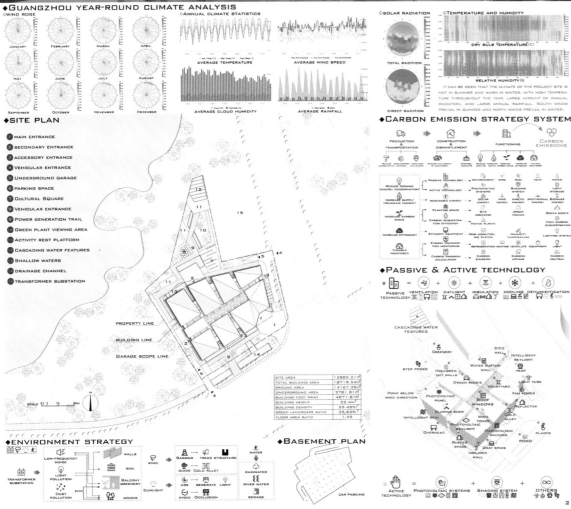

专家点评：
该方案从岭南民居中汲取灵感。将冷巷作为建筑核心交通轴线，结合彩色太阳能玻璃的光影效果，提升了空间的丰富度。热缓冲空间与表皮设计考虑遮阳和隔热，以适应当地气候特点。但是，应优化外部环境与内部空间的连接，降低立面封闭性，提升使用者体验。

Expert Commentary:
The scheme draws inspiration from traditional Lingnan folk houses. The central traffic axis of the building is formed by cold lanes, while the vibrant solar glass curtains enhance the spatial richness through their colorful lighting effects. Both shading and heat insulation are considered in the thermal buffer space as well as the architectural envelope to adapt to local climatic conditions. However, optimizing the connection between the external environment and internal space is necessary to reduce facade closure and improve user experience.

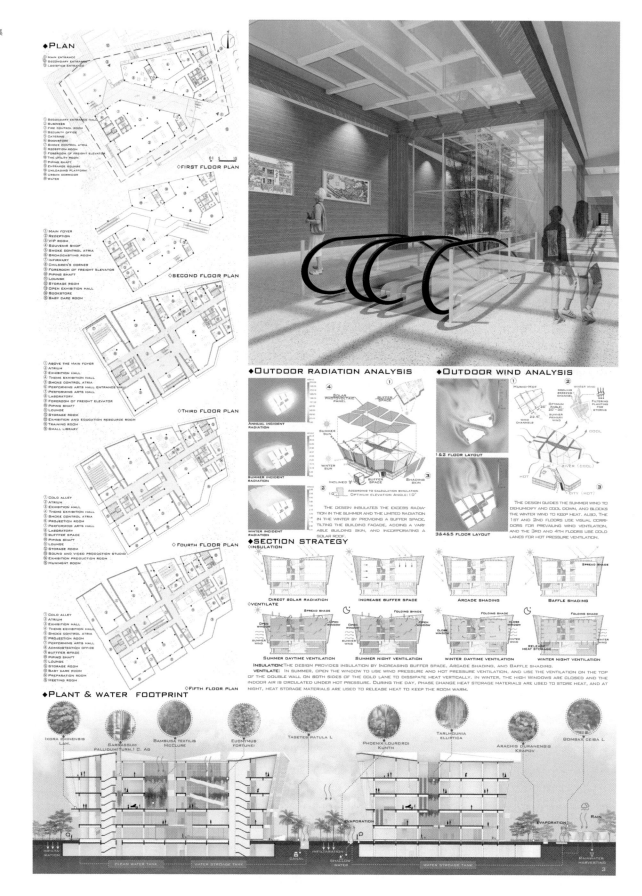

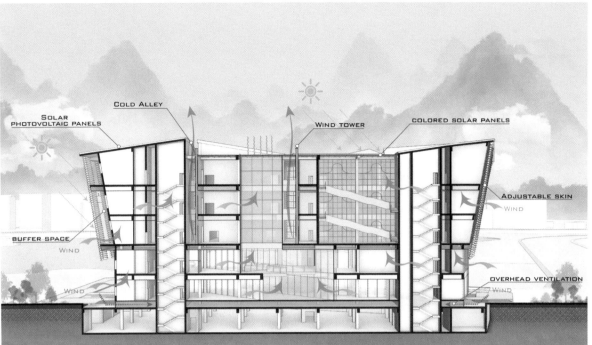

◆ Explosive Structure & Material

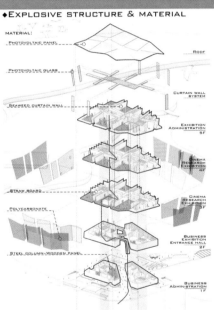

◆ Constructed Specification

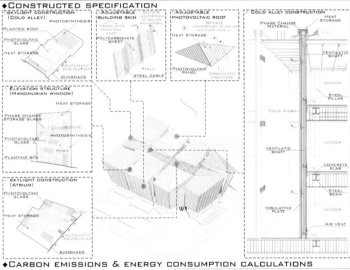

◆ Carbon Emissions & Energy Consumption Calculations

◆ Elevation Plan

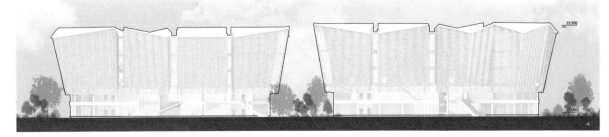

综合奖・三等奖・零碳设计项目
Comprehensive Awards・Third Prize・Zero-Carbon Design Project

注册号：101650
Register Number：101650

项目名称：风・巷・塔
Entry Title：Wind, Alley, Tower

作者：李文钰、李金琦、林永康
Authors：Wenyu Li, Jinqi Li and Yongkang Lin

作者单位：南京工业大学
Authors from：Nanjing Tech University

指导教师：舒欣、吕明扬
Tutors：Xin Shu and Mingyang Lü

指导教师单位：南京工业大学
Tutors from：Nanjing Tech University

■ DESIGN DESCRIPTION

本案位于广州增城科教城内，服务整个科教城及城市的需要。项目主题为：风・巷・塔，借鉴岭南地区的传统建筑布局，以冷巷、拔风井等元素为主要的被动式设计理念，配合以雨水收集、热量处理太阳能BI-PV\BAPV、垂直绿化、双层玻璃幕墙等主动式绿色建筑技术，构筑了一个充分集成绿色、节能、低碳技术和传统元素理念的科技馆。建筑整体流线清晰、功能明确，能够较好地满足游客、周边师生、居民、科研人员的使用需求，同时能够作为一个绿色、零碳建筑的展示馆和体验容器，推广具有地域特色的绿色建筑。

This project is located in Guangzhou Zengcheng Science and Education City, serving the needs of the entire Science and Education City and the city. The theme of the project is: Wind-Alley-Tower. Drawing on the layout of traditional buildings in the Lingnan region, the main passive design concepts include cold alleys and wind extraction shafts, together with active green building technologies such as rainwater collection, heat treatment solar BIPV\BAPV, vertical greening, double-layer glass curtain wall, etc., the project constructs a science and technology museum that fully integrates the concepts of green, energy-saving, low-carbon technologies and traditional elements. The overall flow of the building is clear and functional, which can better meet the needs of tourists, neighboring teachers and students, residents and researchers, and at the same time can serve as a green, zero-carbon building exhibition hall and experience container to promote green buildings with regional characteristics.

■ ECONOMIC ANALYSIS

Site Area: 11447 ㎡
Building Density: 38.6%
Floor Area Ratio: 1.58
Building Area: 18860.1 ㎡
Building Height: 26.5m
Greening Rate: 28%

■ ENVIRONMENTAL ANALYSIS

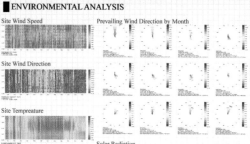

Site Wind Speed | Prevailing Wind Direction by Month | Year-round Prevailing Winds | Dry Bulb Temperature
Site Wind Direction | Enthalpy and Humidity Diagram | Dew Point Temperature
Site Temperature | Solar Radiation | Solar Insolation Radiation
Site Humidity

■ SITE PLAN 1:1000

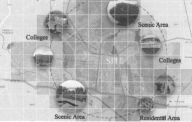

■ SITE ANALYSIS

Colleges | Scenic Area | SITE | Colleges | Scenic Area | Residential Area

■ CULTURAL ANALYSIS

South of the five ridges — The Lingnan immigrants who moved their families to the south brought with them new forms of architecture, making Lingnan architecture exquisite and chic. Commercial and Western architecture was prevalent.

209 B.C. | The Last Years Of the Song Dynasty | The Ming (1368-1644) and Qing (1644-1911) Dynasty | Republic of China (1912-1949) | Now

■ CONCEPT GENERATION

TRADITIONAL
High walls, cold alleys and patio. | Ground floor elevation. | Wind catching tower. | High mountain walls of the wok house.

CONTEMPORARY
Architectural form organization. Continuation of the traditional layout. | Modern Wind catching tower. | Double glass curtain wall and vertical greening.

Heritage & Innovation
Inherit the traditional methods, combined with new materials and technologies, inheritance and innovation in tradition and de-homogenization. This is a way out for regional green building.

■ BLOCK GENERATION

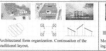

1. Original site.(Zengcheng,Guangzhou)
2. Lift the block at the desired height.
3. Block division.
4. Slope of the roof.
5. Setting up a rooftop garden.
6. Setting up four chimneys.
7. Setting up shading elements and vertical greening.
8. Refinement of architectural details.

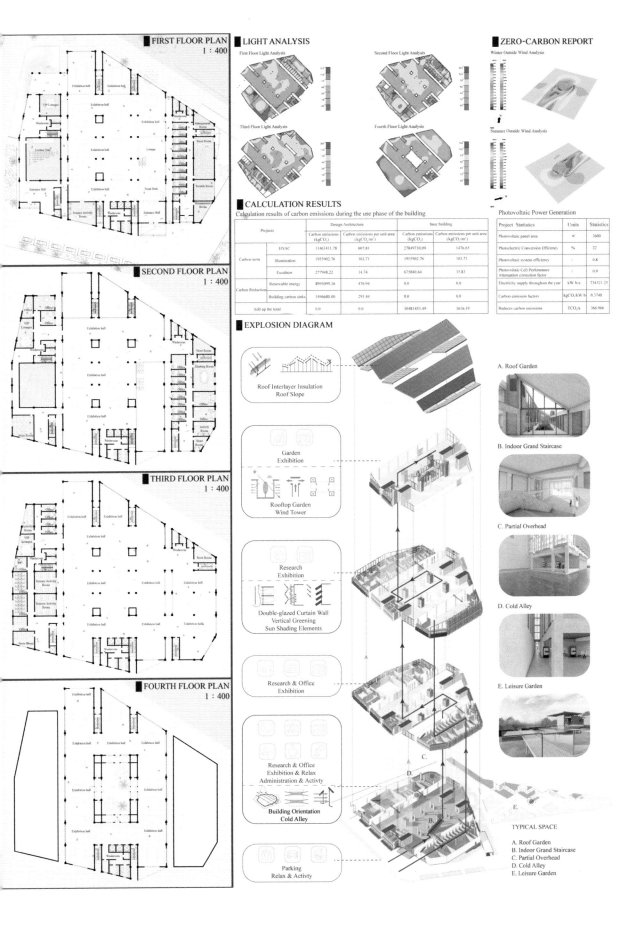

专家点评:

该方案节能技术应用完善。拔风井的内表面使用了降温材料,同时增加了防火材料和防火卷帘,为场馆带来更好通风效果的同时增加了安全性,但作为科技馆的设计方案,需加强公共建筑在场地、立面、空间等方面特性的考量。

Expert Commentary:

A comprehensive application of energy-saving technologies has been effectively demonstrated in the scheme. Cooling materials are utilized on the inner surface of the ventilation shaft, while fire-resistant materials and fire-resistant rolling curtains are incorporated to enhance venue ventilation and ensure safety. However, as a design scheme of a science and technology museum, it is essential to further consider its public architectural characteristics in terms of site, facade and space.

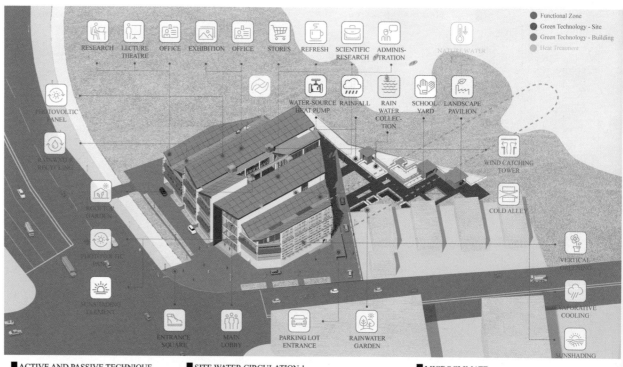

ACTIVE AND PASSIVE TECHNIQUE

SITE WATER CIRCULATION 1
RAINWATER COLLECTION & EVAPORATION

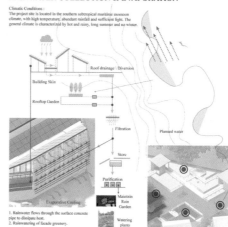

Climatic Conditions:
The project site is located in the southern subtropical maritime monsoon climate, with high temperature, abundant rainfall and sufficient light. The general climate is characterized by hot and rainy, long summer and no winter.

1. Rainwater flows through the surface concrete pipe to dissipate heat.
2. Rainwatering of facade greenery.

The site water recycling process is one of the most important means of green site design by linking building facades, roofs, and site rain gardens to collect rainwater and utilize it for irrigation, building surface cooling, and other processes after filtering, storing, and purifying it.

MICROCLIMATE

A. SUNSHADING — Use of facade shading elements and vertical greening to reduce heat convection and increase evaporation for cooling.

B. NATURAL OXYGEN BAR — Rooftop garden planting and greening of the building site constitute a natural oxygen bar.

C. COLD ALLEY: AIR FLOW — Building form organization, site design, forming cold alleys to promote air circulation.

D. WATER RECYCLING — Building water recycling system utilizing water cooling, rainwater harvesting, evaporation and heat dissipation.

E. COOLING — Reduction of radiation using highly diffuse paving on building surfaces.

CUSTOMER BASE ANALYSIS

SITE WATER CIRCULATION 2
COOLING FROM NATURE WATER SOURCES

Utilizing the characteristics of constant temperature and coldness of natural water sources, laying cooling water pipes in the building to cool down the building, solving the problem of heat conversion and processing in the building.

BAPV (Building Attached Photovoltaic)

PV ROOF — Advantage: generate electricity, reduce heat entering the room

SOLAR VISOR — Advantage: generate electricity, reduce direct sunlight

CONCENTRATED SOLAR POWER — Advantage: generate electricity, reduce energy consumption

PV CANOPY — Advantage: generate electricity, enrich the site, provide a place to rest

ELEVATION PLAN 1:400

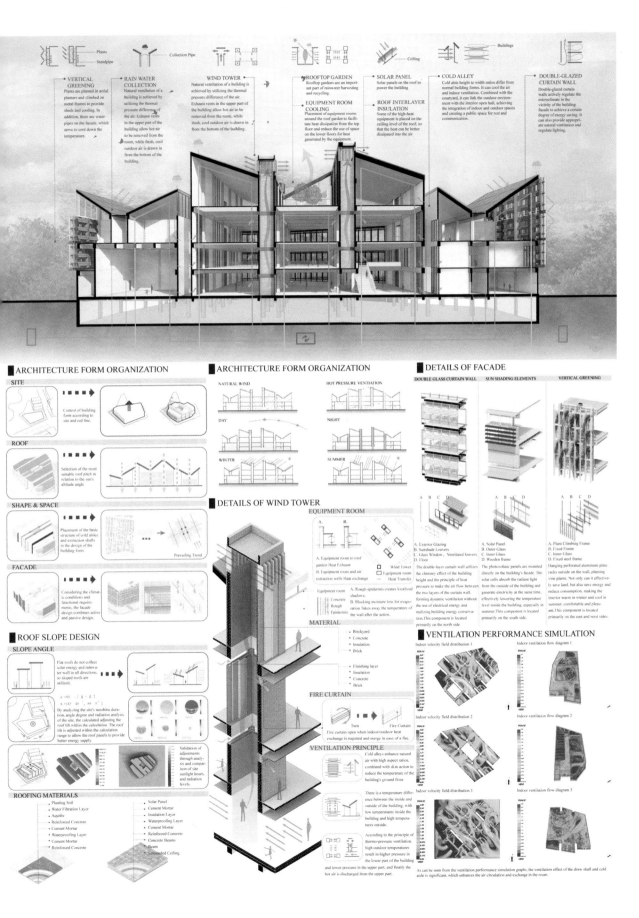

综合奖·三等奖·零碳设计项目
Comprehensive Awards · Third Prize · Zero-Carbon Design Project

注册号：101731
Register Number：101731

项目名称：乘风好去
Entry Title：Wind Catcher

作者：邢惠玥、石伊、滕震东
Authors：Huiyue Xing, Yi Shi and Zhendong Teng

作者单位：南京工业大学
Authors from：Nanjing Tech University

指导教师：吕明扬、罗靖
Tutors：Mingyang Lyu and Jing Luo

指导教师单位：南京工业大学
Tutors from：Nanjing Tech University

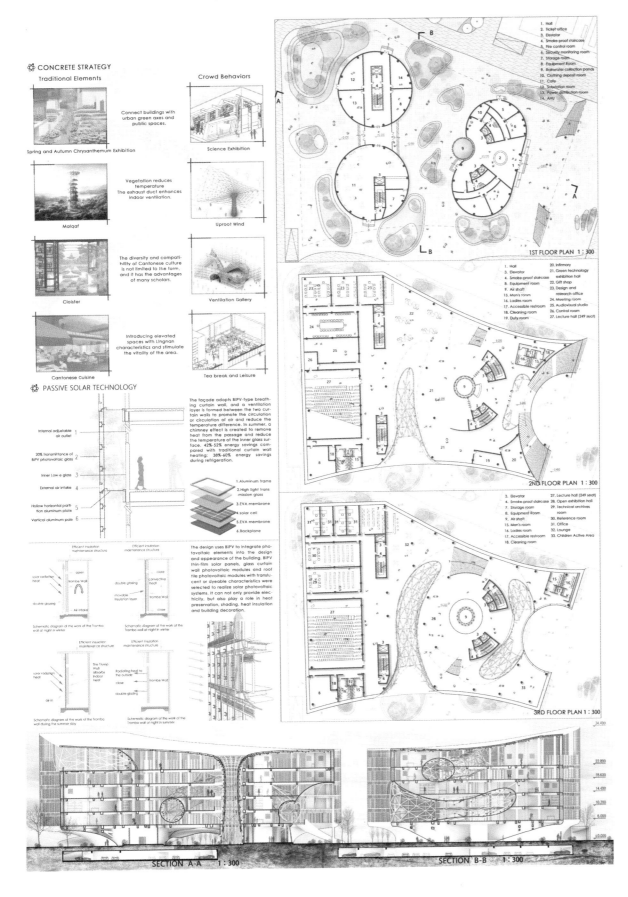

专家点评：

该方案紧密结合周边场地的情况，较好地呼应了周边的城市道路与景观环境。底层架空的形式和通风井的设置可以有效提升建筑通风条件，适应当地气候。对内部管状风道的设计丰富了空间效果，但其曲线的形态不利于保持进入建筑的风速。同时，复杂的结构设计和相关材料的选择不利于降低建筑全生命周期的碳排放。

Expert Commentary:

The scheme seamlessly integrates the site environment into the building design, effectively responding to surrounding urban transportation and landscaping conditions. The elevated ground floor and strategically placed ventilation shafts significantly enhance ventilation capabilities and adaptability to local climatic characteristics. Despite the enhanced spatial effect and visual interest produced by the design of an internal tubular air duct, its curved shape hinders the maintenance of optimal wind speed upon entry into the building. Additionally, both the complex structural design and choice of building materials fail to contribute favorably towards reducing carbon emissions throughout the entire life cycle of the building.

ARCHITECTURAL DETAIL RENDERINGS

THE DESIGN OF THE BUILDING ENTRANCE USES A CIRCULAR CUT VOLUME TO FORM AN ENTRANCE WITH A LINE OF SIGHT GUIDING EFFECT, AND THE CUT SHAPE OF THE SPACE BUILDING FORMS AN EXTERNAL VOID FULL OF POWER, SO THAT THE INTERIOR AND EXTERIOR SPACE ARE CONNECTED, WHICH CAN SHADE THE SUN AND RAIN, ALLOW NATURAL LIGHT TO ENTER, AND GUIDE THE FLOW OF PEOPLE. AT THE SAME TIME, THE AIR DUCT IS CUT AS A VENTILATION PORT.

HERE IS THE TRAFFIC SPACE FROM THE 2ND FLOOR TO THE 3RD FLOOR, WHICH CONFORMS TO THE UPWARD TREND OF THE WIND DUCT, FORMS A NATURAL LINE OF SIGHT GUIDANCE, CREATES A VIVID AND INTERESTING TRAFFIC SPACE, AND REDUCES THE OVERALL SENSE OF THE PLANE.

THIS IS A FOUR-STORY OUTDOOR PLATFORM, AT THE INTERSECTION OF THE AIR DUCT AND THE NORTH FAÇADE, MIXED AND USED AS A WIND-TO-EXCHANGE AIR OUTLET AS A BUFFER SPACE BETWEEN THE BUILDING AND THE EXTERNAL ENVIRONMENT, NATURALLY FORMING A VIEWING PLATFORM, PROVIDING SEMI-OUTDOOR RESIDENCE SPACE FOR USERS OF THE UPPER SPACE. AT THE SAME TIME, IT WEAKENS THE PHYSICAL SENSE OF THE FORM, MAKING THE SCIENCE AND TECHNOLOGY MUSEUM LOOK LIGHTER.

THE BUILDING RAISES THE LAKE SURFACE TO SHAPE A HIGH-SCALE VIEW OF THE LAKE, AND AT THE SAME TIME FOR PEOPLE TO SEE THE BUILDING FROM THE LAKE, AND THE WELL-DESIGNED EXTERIOR SHAPE MAKES IT A LANDMARK BUILDING. THE ROOF IS GRADUALLY RAISED FROM EAST TO WEST, SO THAT THE TOP FLOOR FORMS AN INDOOR SPACE THAT GRADUALLY RISES FROM EAST TO WEST, FORMING AN INDOOR SPACE WITH THE ROLE OF SIGHT GUIDANCE.

DATA SIMULATION

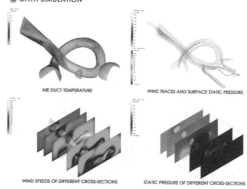

AIR DUCT TEMPERATURE

WIND TRACES AND SURFACE STATIC PRESSURE

WIND SPEEDS OF DIFFERENT CROSS-SECTIONS

STATIC PRESSURE OF DIFFERENT CROSS-SECTIONS

功能剖面 FUNCTIONAL PROFILE

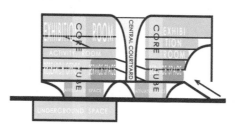

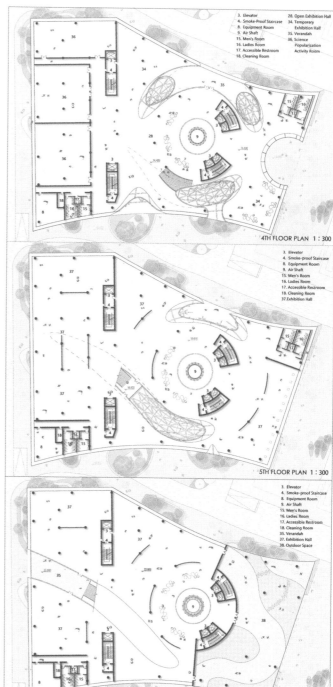

4TH FLOOR PLAN 1:300

5TH FLOOR PLAN 1:300

6TH FLOOR PLAN 1:300

NORTH ELEVATION 1:300

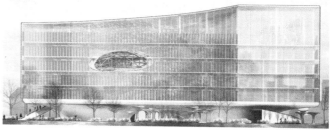

ENERGY SAVING ANALYSIS

In summer, the zenith angle of the sun on the summer solstice in Guangzhou is 89°64', and sunlight is used to reflect and scatter light indoors. The solar panels on the west side and part of the south side provide shading power generation and form a good shadow effect at the same time.

In winter, the zenith angle of the sun on the winter solstice in Guangzhou is 43°64', and the air space and double-glazed windows achieve good thermal insulation effect. The layout of the south-facing corridor forms a good thermal insulation buffer space and realizes winter insulation. And through the different sizes of vents, block the north wind in winter.

DATA ON CARBON EMISSIONS FROM BUILDING OPERATIONS

SITE INSTALLATION AND WATER CYCLE ANALYSIS

The solar energy devices on the site will be installed with solar panels from multiple angles, striving for the optimal use of solar energy. They are used to provide charging plugs, WiFi routers, and information screens.

The rainwater collection device on the site will purify the water source and collect rainwater, which can be used for irrigation, drinking and fire hydrant water.

STRUCTURAL EXPLOSION DIAGRAM

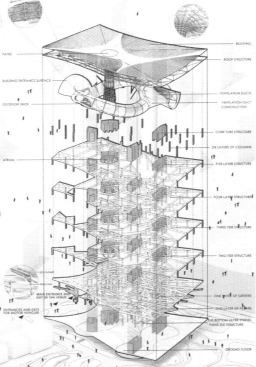

NET-ZERO CARBON STRATEGY

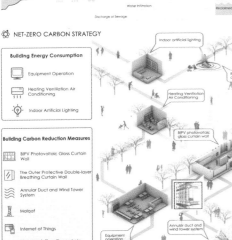

Building Energy Consumption
- Equipment Operation
- Heating Ventilation Air Conditioning
- Indoor Artificial Lighting

Building Carbon Reduction Measures
- BIPV Photovoltaic Glass Curtain Wall
- The Outer Protective Double-layer Breathing Curtain Wall
- Annular Duct and Wind Tower System
- Malqaf
- Internet of Things
- Monocrystalline silicon photovoltaic panels are installed on the roof and site
- Rainwater Collection System
- Uproot Wind

综合奖·三等奖·零碳设计项目
Comprehensive Awards·
Third Prize·Zero-Carbon
Design Project

注册号：101804
Register Number：101804

项目名称：光盒作用
Entry Title：Photosynthesis

作者：马子雯、曹高源、冯康芸、
　　　卞雨馨、许天一
Authors：Ziwen Ma, Gaoyuan Cao,
　　　Kangyun Feng, Yuxin Bian and
　　　Tianyi Xu

作者单位：天津大学
Authors from：Tianjin University

指导教师：朱丽
Tutor：Li Zhu

指导教师单位：天津大学
Tutor from：Tianjin University

■ Design Desription

本方案以"光伏树"和"共生盒"为线索，以零碳技术为手段，采用光伏建筑一体化设计、雨水收集系统等主、被动式太阳能技术以及适宜的低碳技术，生成一个独特的创新和实验的空间连续体，实现建筑碳排放为零。在合理规划布局的基础上，通过中庭和屋顶花园等设计弱化建筑规模，将建筑从城市群缓慢向公园和水系过渡。在建筑单体设计中，以具有张力的造型实现大型技术装置集成绿色低碳技术和完整表皮负荷等特点，使技术装置与建筑表皮有机结合。

This design takes "photovoltaic tree" and "symbiotic box" as the clue, uses zero-carbon technology as the means, adopts photovoltaic building integrated design, rainwater collection system and other active and passive solar technology as well as appropriate low-carbon technology, generates a unique spatial continuum of innovation and experiment, and realizes zero carbon emission of buildings. On the basis of reasonable planning and layout, the design of atrium and roof garden weakens the scale of the building, and slowly transitions the building from the urban agglomeration to the park and lake. In the design of building monomer, the tension shape is used to realize the characteristics of integrating green and low carbon technology and complete skin load, so that the technical device and the building skin can be organically combined.

■ Site Analysis

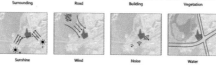

■ Climate Analysis

The project is located in the subtropical maritime monsoon climate, with high temperature, abundant rainfall and sufficient sunlight.

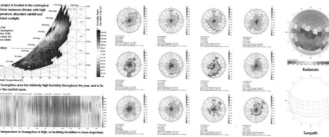

The temperature in Guangzhou is high, so building insulation is more important.

■ Site Plan 1:1000

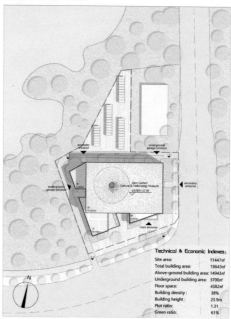

Technical & Economic Indexes:
Site area: 11447㎡
Total building area: 18643㎡
Above-ground building area: 14943㎡
Underground building area: 3700㎡
Floor space: 4382㎡
Building density: 38%
Building height: 23.9m
Plot ratio: 1.31
Green ratio: 61%

■ Diagram of Design Process

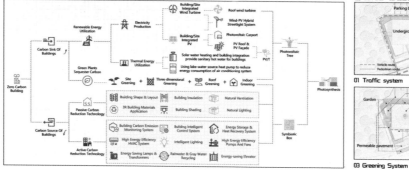

■ Site Design

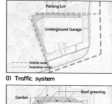

01 Traffic system　02 Function Layout
03 Greening System　04 Energy System

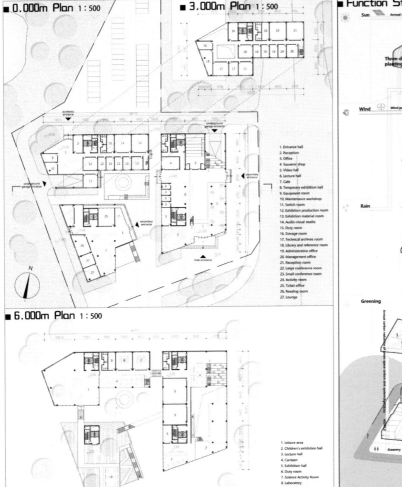

专家点评：
该方案的设计充分考虑了建筑作为公共建筑的特性和场地的影响，空间高差与流线设计多样，入口空间活跃丰富。主动、被动技术整体性好，但需优化通风井形态以提升通风效率，并深入考虑屋顶结构与遮阳、绿化的经济性。

Expert Commentary:
The scheme fully considers the characteristics of a public building and its impact on the site environment. It features differentiated spatial heights and various circulations, creating vibrant and dynamic entrance spaces. The design effectively integrates both active and passive technologies; however, it is necessary to optimize the shape of the ventilation shaft to enhance ventilation efficiency. Additionally, further consideration should be given to the roof structure as well as the cost-efficiency for shading and landscaping.

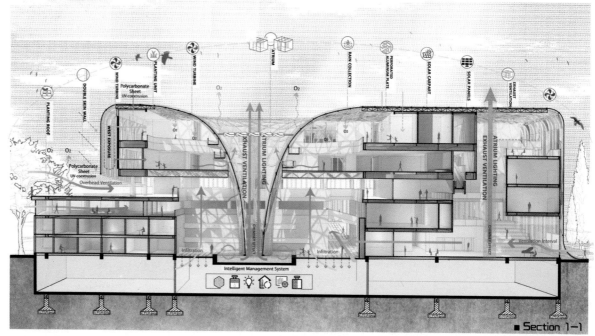
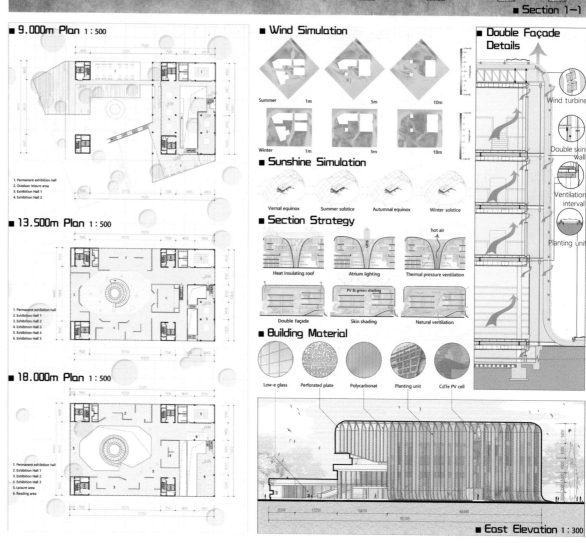

Function And Structure Schematic

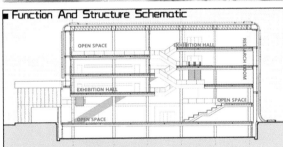

Carbon Emission Monitoring & Intelligent Control System

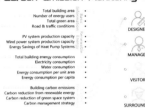

Carbon Emission Calculation

Zero Carbon Guilding Technology Strategy

Multi-dimensional Green Plant Carbon Sink
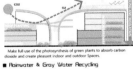
Make full use of the photosynthesis of green plants to absorb carbon dioxide and create pleasant indoor and outdoor Spaces.

Rainwater & Gray Water Recycling
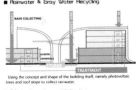
Using the concept and shape of the building itself, namely photovoltaic trees and roof slope to collect rainwater.

Waste Recycling and Utilization
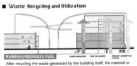
After recycling the waste generated by the building itself, the material or energy is reused for the building.

Building integrated energy systems system
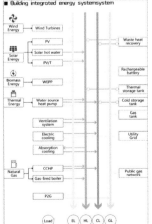

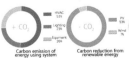
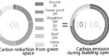

综合奖·三等奖·零碳提升项目
Comprehensive Awards · Third Prize · Zero-Carbon Promotion Project

注册号：101441
Register Number：101441

项目名称：风光相伴，绿意同生
Entry Title：Breeze and Sunshine Accompany, Greenery and Architecture Coexist

作者：许轩宁、梁中杰、洪宇倩、谭淏蓝
Authors：Xuanning Xu, Zhongjie Liang, Yuqian Hong and Haolan Tan

作者单位：中国矿业大学
Authors from：China University of Mining and Technology

指导教师：段忠诚、姚刚、邵泽彪、马全明
Tutors：Zhongcheng Duan, Gang Yao, Zebiao Shao and Quanming Ma

指导教师单位：中国矿业大学
Tutors from：China University of Mining and Technology

General Plan | General plane of building after renovation

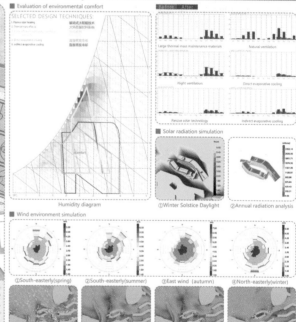

Climate Analysis | Building sunshine, radiation, humidity & wind speed

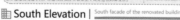

South Elevation | South facade of the renovated building

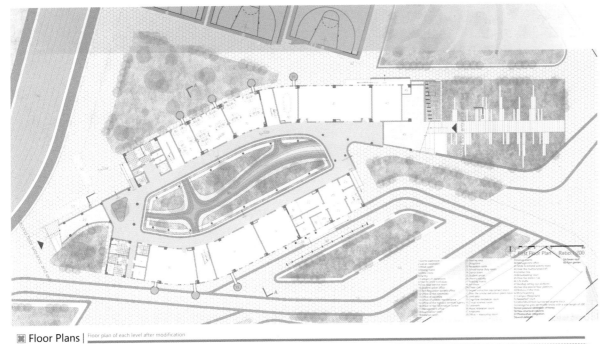

专家点评：
方案将光伏与建筑造型融合，雨水花园设计精心，热压烟囱与导风板设置改善了通风与光环境，但地源热泵使用考虑不足，应结合广州气候深入考量。整体而言，方案创新实用，但需进一步完善细节。

Expert Commentary:
The scheme successfully integrates photovoltaic technologies into the architectural form, meticulously designs the roof garden for rainwater collection, and incorporates thermal ventilation chimneys and wind deflectors to enhance the ventilation and lighting environment. However, further consideration should be given to the utilization of ground source heat pumps based on the local climate in the Guangzhou area. Overall, the scheme is innovative and practical but requires further refinement at a detailed level.

Floor Plans | Floor plan of each level after modification

North Elevation | North facade of the renovated building

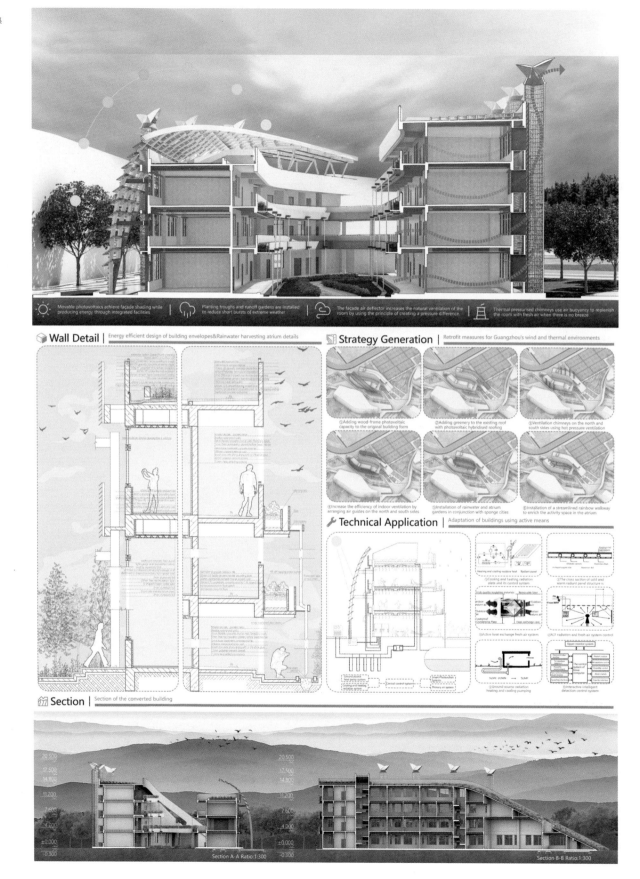

🗺 Strategy Analysis | Photovoltaic panel angle optimization, thermal ventilation efficiency, indoor lighting simulation & carbon emission calculation

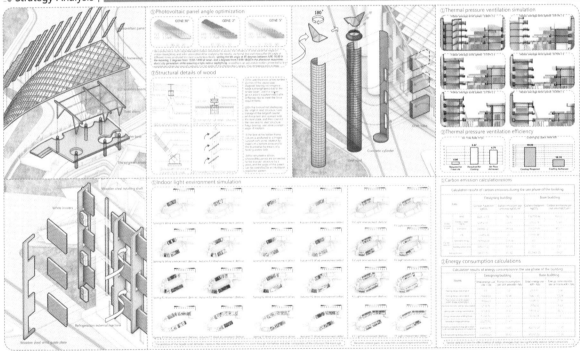

🏙 Scene Perspective | Facade wind deflectors, hot-pressed chimneys, rainwater gardens & production capacity photovoltaic panels

综合奖・优秀奖・零碳设计项目
Comprehensive Awards · Honorable Mention · Zero-Carbon Design Project

注册号：101309
Register Number：101309

项目名称：零度科创馆・岭
Entry Title：Lingnan Zero Carbon Science and Technology Innovation Museum

作者：汪郑政、高望皓、项涛、赵文祥、武泽宇、翁得玛克・库晚、韩世仓、许文博、吉南木・木沙、乃依马・麦麦提艾则孜
Authors：Zhengzheng Wang, Wanghao Gao, Tao Xiang, Wenxiang Zhao, Zeyu Wu, Wengdemake. Kuwan, Shicang Han, Wenbo Xu, Jinanmu.Musha and Naiyima.Maimaitiaizezi

作者单位：新疆大学
Authors from：Xinjiang University

指导教师：樊辉、艾斯卡尔・模拉克、陈善婷
Tutors：Hui Fan, Aisikaer.Molake and Shanting Chen

指导教师单位：新疆大学
Tutors from：Xinjiang University

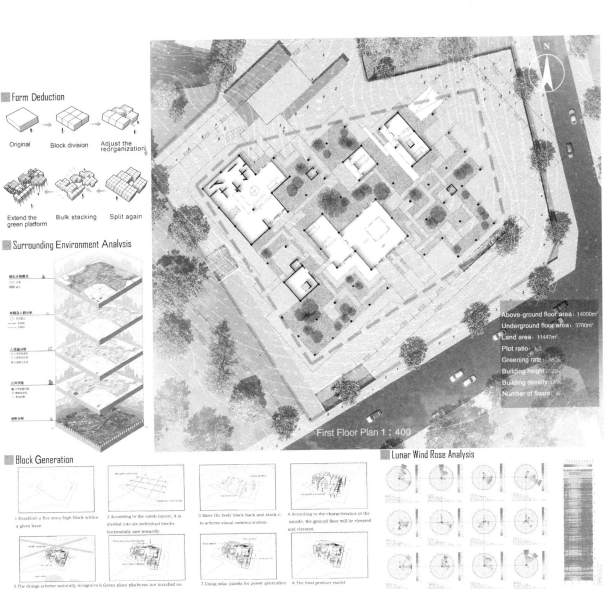

Form Deduction

Original → Block division → Adjust the reorganization → Extend the green platform → Bulk stacking → Split again

Surrounding Environment Analysis

Above-ground floor area: 14000m²
Underground floor area: 3700m²
Land area: 11447m²
Plot ratio: 1.2
Greening rate: 38%
Building height: 29
Building density: 22%
Number of floors: 6

First Floor Plan 1:400

Block Generation

1. Establish a five story high block within a given base.
2. According to the comb layout, it is divided into six individual blocks horizontally and inwardly.
3. Move the body block back and stack it to achieve visual communication.
4. According to the characteristics of the arcade, the ground floor will be elevated and elevated.
5. The design scheme naturally integrates with the surrounding site environment.
6. Green plant platforms are installed on the roof to create a three-dimensional landscape to reduce carbon emissions.
7. Using solar panels for power generation.
8. The final product model.

Lunar Wind Rose Analysis

Cultural and Green Building Analysis

Comb layout:	The 'comb style layout' is a typical village layout form in the Lingnan region of China.	Sunshade:	The 'high walls and narrow alleys' formed by gable walls and narrow cold tunnels use the building's own form to create a large area of shadow.	
Characteristic residential buildings:	A courtyard consisting of a three bay main building with two corridors and a courtyard in front.	Ventilate:	The height to width ratio of the gable is large, and under the action of solar thermal radiation, it forms thermal ventilation in the tunnel. The narrower the tunnel.	
Bamboo tube style residential buildings:	Bamboo tube house is a single bay, narrow width, deep and straight tube shaped residential building.	Heat insulation:	The hollow structure of the empty bucket wall effectively isolates and cushions the heat in the external environment of the wall.	
Arcade:	The elevated gray space of the eaves corridor on the bottom floor provides convenience for pedestrians on the street.	Moisture-proof:	By cleverly setting the height of the building eaves to attract light into the room, the problem of spring flooding is solved.	

Enthalpy Hygrogram Analysis

Rainfall Analysis

Design Instructions

项目位于广州市增城区，属于亚热带季风气候。该项目紧密结合当地岭南文化，通过横巷和里巷的肌理关系将空间分为六大板块，并根据现场周边环境的关系以及岭南园林的特点，将建筑进行退台处理，结合当地气候和风向流向将建筑逐级进行底层架空处理，建筑在技术方面应用了太阳能光伏板、太阳能幕墙、冷巷原理、太阳能折叠双层幕墙结构、导光管、断桥处理、节能光源以及蓄水池等，项目与自然环境融合紧密，有优化的送风排风系统、电力系统、材料碳排指标，最终使项目实现零碳目标。

The project is located in Zengcheng District, Guangzhou, belongs to the subtropical monsoon climate, the project closely combines the local Lingnan culture, through the texture relationship between Hengxiang and Lixiang to divide the space into six major plates, and according to the relationship between the surrounding environment of the site and the characteristics of Lingnan gardens, the building will be treated as a retreat, combined with the local climate and wind direction to gradually carry out the ground floor overhead treatment, the building in terms of technology application of solar photovoltaic panels, solar curtain wall, cold lane principle, solar folding double-layer curtain wall structure, light guide tube, broken bridge treatment, Energy-saving light source, as well as reservoir, etc., the project is closely integrated with the natural environment, and there are optimized air supply and exhaust systems, power systems, and material carbon emission indicators, which ultimately enable the project to achieve the goal of zero carbon.

Second Floor Plan 1:460

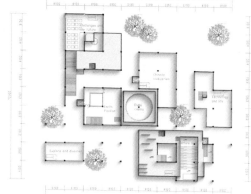

Three-story Floor Plan 1:460

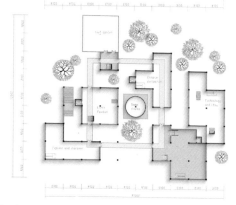

Carbon Cycle System

HVAC systems — The garden landscape introduces water into the superstructure and forms a circulatory system in the form of a water curtain.

Lighting with elevator — Use light guides to bring natural light into the room, while conveying inside and outside, saving light, and using ultra-low energy elevators.

Solar energy — The project is equipped with a photovoltaic curtain wall and a folding double-layer solar panel curtain wall, which makes efficient use of light energy.

Wind energy — In the design process, the inside and cross alleys are arranged diagonally to the wall, and the cold air is introduced through the hot pressure effect, and the hot air is exported.

Construct Node Analysis

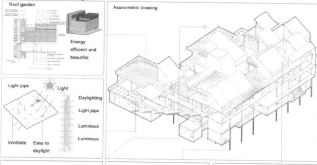

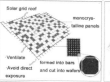 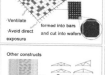 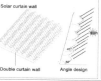 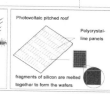

Energy saving instructions

1. The monitoring method of carbon emission data of building operation, and the reduction of carbon emission of equipment system operation through intelligent control.
2. According to the winter solstice sun height Angle and building deviation Angle to determine the solar panel Angle of 45 degrees. According to the cold lane effect, the overall Angle of the curtain wall is maintained at 80 degrees, which can maximize the use of solar energy and avoid direct exposure.
3. The roof garden uses more advanced drip irrigation technology to save energy; The use of cross pillars saves material while using lower carbon materials.

Elevation

Sunshine and Wind Energy Analysis

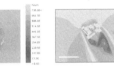

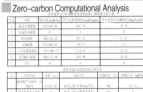

Zero-carbon Computational Analysis

Solid Model

Planning and operation of zero-carbon systems

Analysis of Energy-efficient Materials

Curtain wall solar panels
Some exterior walls use photovoltaic glass curtain walls for power generation. The metal halide perovskite thin film layer is used to convert solar energy into electrical energy, and the positive terminal and negative terminal are used to output the electrical energy.

Bright tile sunshade roof
Part of the shading is made of traditional material shingles. When inlaying bright tiles, thin bamboo sheets should be woven into a grid first, and then the tiles should be embedded in it to achieve good lighting effects.

Green brick cladding
Some walls are covered with Lingnan blue bricks on the outside. The wall is recessed in a hollow space with blue bricks, which have the effect of sound insulation and cooling. The blue brick wall often can withstand collapse for a hundred years.

Traditional oyster shell wall
The exterior walls of the Central Theater are made of traditional oyster shells. In cloudy and rainy weather, rainwater will quickly flow along the oyster shell to keep the indoor dry and tidy. The arrangement of pieces with burrs at a certain gap, which is beneficial for isolating.

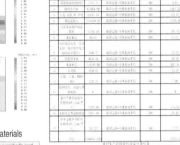

Sectional Perspective

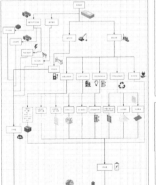

camphor tree · water lily · Mizailan · Cosmos · Acacia · Oil fir

Landscape and Natural Ventilation Analysis

Magnolia · palm · Golden Rong · Money grass · calamus · Ficus microphylla

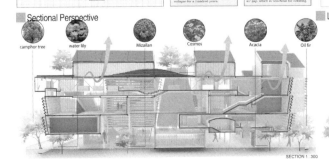

SECTION 1:300

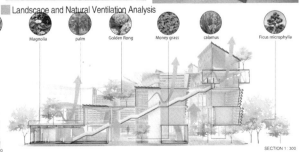

SECTION 1:300

综合奖·优秀奖·零碳设计项目
Comprehensive Awards · Honorable Mention · Zero-Carbon Design Project

注册号：101358
Register Number：101358

项目名称：拾级叠院
Entry Title：Clambered Up-Yard

作者：卫茹冰、赵子瑶、陈东晓、孔令熙、王垚橙、舒琨狄
Authors：Rubing Wei, Ziyao Zhao, Dongxiao Chen, Lingxi Kong, Yaocheng Wang and Kundi Shu

作者单位：俄罗斯乌拉尔联邦大学、俄罗斯圣彼得堡彼得大帝理工大学、北京建筑大学、华北水利水电大学乌拉尔学院、天津大学

Authors From：Peter the Great St. Petersburg Polytechnic University（Уральский федеральный университет имени первого Президента России Б. Н. ЕльцинаУральский федеральный университет）, Peter the Great St. Petersburg Polytechnic University（Санкт-Петербургский политехнический университет Петра Великого）, Beijing University of Civil Engineering and Architecture, Architecture Dept. Ural Institute, North China University of Water Resources and Electric Power, and Tianjin University

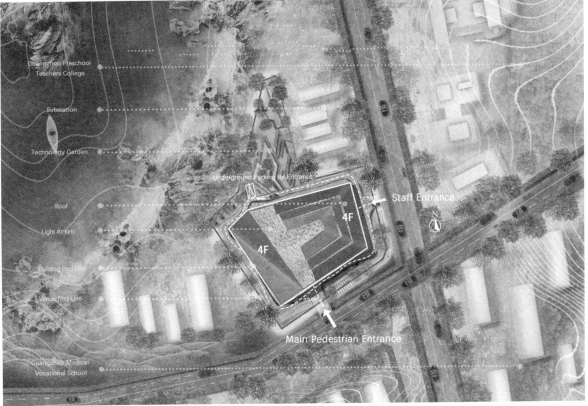

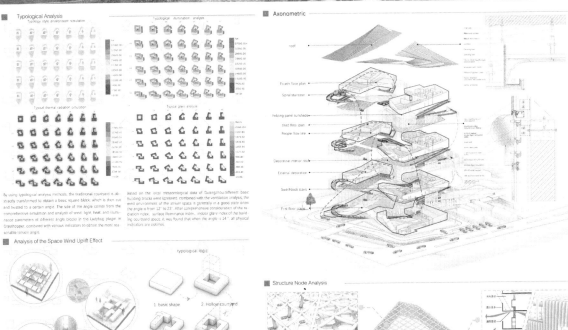

指导教师：尼基蒂娜·娜塔莉亚·帕夫洛芙娜、塔蒂亚娜·列昂尼多夫娜·西曼金娜、李春青

Tutors：Никитина Наталия Павловна, Симанкина Татьяна Леонидовна and Qingchun Li

指导教师单位：俄罗斯乌拉尔联邦大学、俄罗斯圣彼得堡彼得大帝理工大学、北京建筑大学

Tutors from：Peter the Great St.Petersburg Polytechnic University (Уральский федеральный университет имени первого Президента России Б.Н. ЕльцинаУральский федеральный университет), Peter the Great St.Petersburg Polytechnic University (Санкт-Петербургский политехнический университет Петра Великого) and Beijing University of Civil Engineering and Architecture

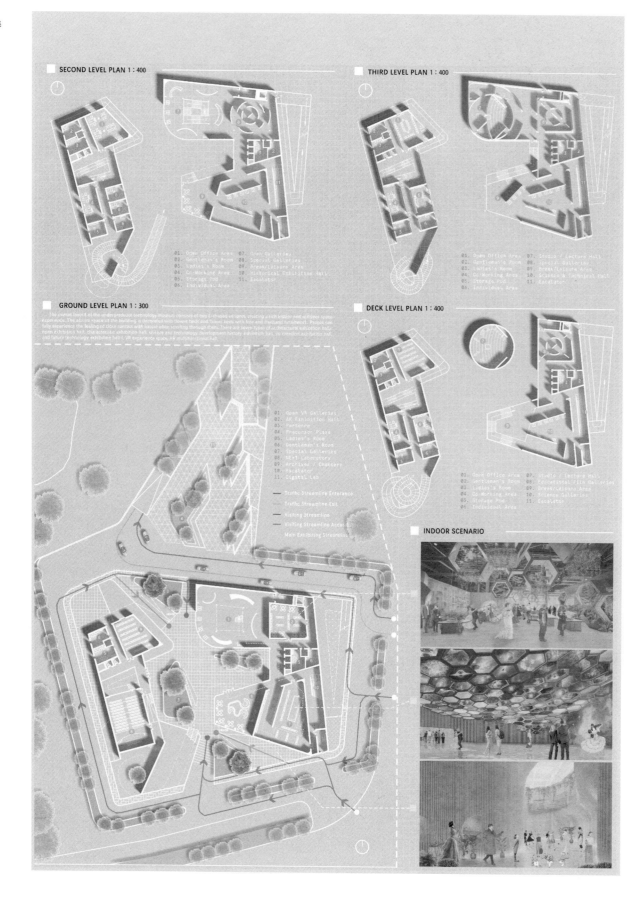

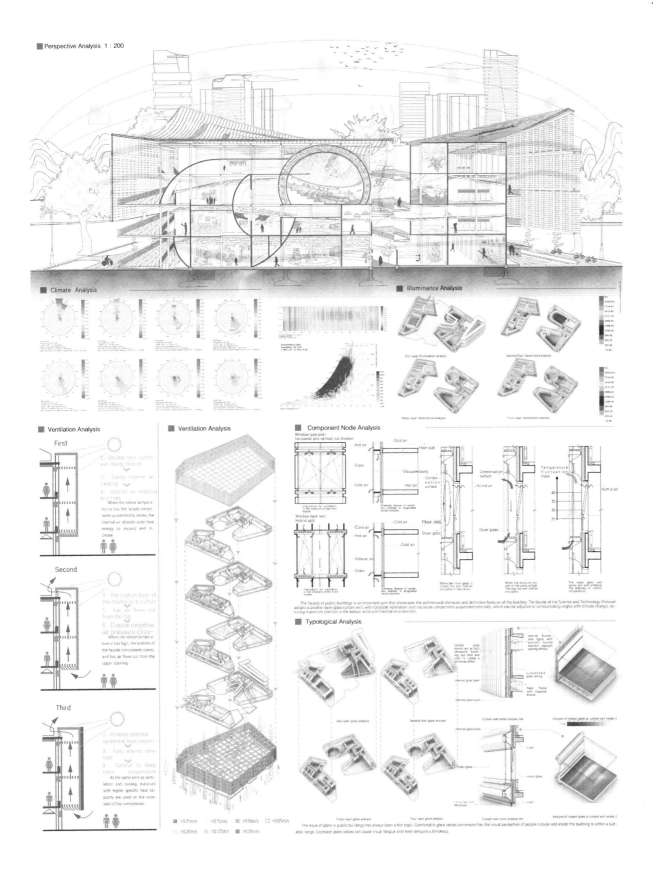

综合奖・优秀奖・零碳设计项目
Comprehensive Awards · Honorable Mention · Zero-Carbon Design Project

注册号：101363
Register Number：101363

项目名称：绿光新谷
Entry Title：Green Valley

作者：孟圆、孙怡文、陈诗扬、杨一涵、马超毅、陈胤企
Authors：Yuan Meng, Yiwen Sun, Shiyang Chen, Yihan Yang, Chaoyi Ma and Yinqi Chen

作者单位：长安大学、东北林业大学、阿德莱德大学
Authors from：Chang'an University, Northeast Forestry University and University of Adelaide

指导教师：夏博、崔鹏
Tutors：Bo Xia and Peng Cui

指导教师单位：长安大学、东北林业大学
Tutors from：Chang'an University, Northeast Forestry University

01 Base Analysis

02 Location Analysis

03 Humanistic Analysis

04 Application Analysis

05 Technical Analysis

06 Model Display

07 Design Specification

"绿光新谷"是指通过利用绿色技术和光照特点，结合广州地区传统建筑中"风巷""采光井"的设计原理和设计语言，着重考虑对于风的利用和隔热降温的设计策略，巧妙实现科技建筑的被动式节能，使得建筑物在其整个生命周期内都能保持低碳、高效和环保。建筑结合场地周围环境，以教育功能为主，为科教城以及城市提供文化交流、技术展示等功能，同时充分集成绿色、节能、低碳技术，成为诠释绿色低碳理念的城市窗口。

"Green Light New Valley" means that through the use of green technology and lighting characteristics, combined with the design principle and design language of "wind alley" and "light well" in traditional buildings in Guangzhou, the design strategy of wind utilization and heat insulation and cooling is emphasized, and the passive energy saving of the science and Technology museum building is cleverly realized, so that the building can maintain low carbon, high efficiency and environmental protection throughout its life cycle. Combined with the surrounding environment of the site, the building focuses on educational functions, providing cultural exchange and technology display for the science and education city and the city, while fully integrating green, energy-saving and low-carbon technologies, becoming an urban window to interpret the concept of green and low-carbon.

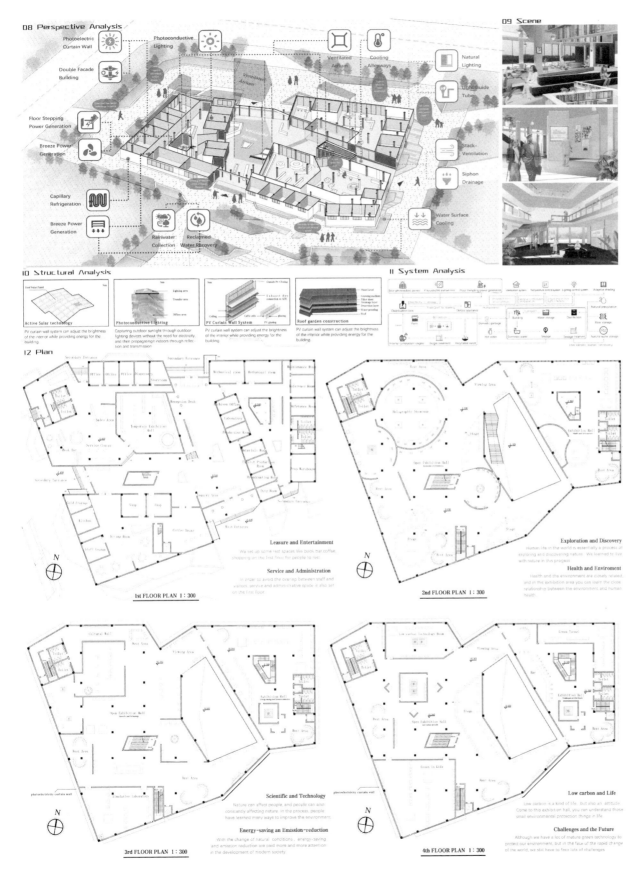

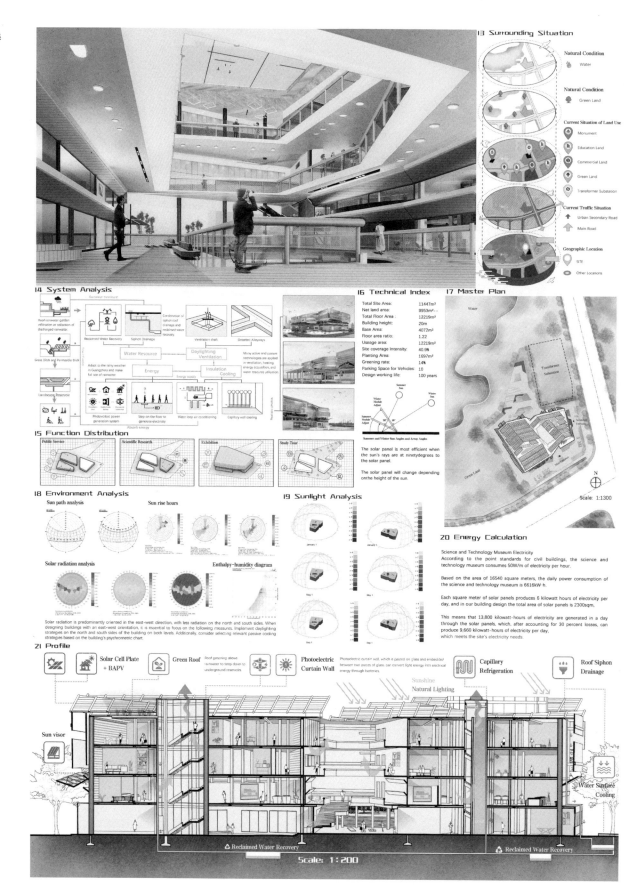

22 Block Generation

Site Scale
The base area is 4072 square meters.
The east and south sides of the site are main roads, and the northwest side is water.

Functional District
The building mainly consists of three major functions: exhibition, public service, and scientific research.

Vehicle Streamline
The south side of the land is the main urban road, the east side is secondary urban road. Therefore, the parking entrance and the drop off area are located on the east side.

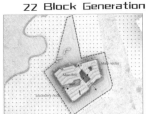
Main Axis
The building consists of a central main axis and secondary axis facing the wetland park, with the main node being the central atrium of the building.

Pedestrain Streamline
The external streamline of the building mainly forms a loop around the building and outdoor exhibitions, while the internal streamline revolves around the exhibition hall.

Entrance Setting
The south side is the main entrance, with two secondary entrances and there is an entrance for office staff.

Airflow Settings
The dominant wind direction of the site is the southeast wind, and the secondary wind direction is the northeast wind. We have set up air intakes on the windward side.

Landscape Resources
The site is rich in natural resources, with a large area of water and green space on the northwest side. The site is equipped with landscape nodes and vertical greening.

23 Carbon Emission Calculation Report

$Y = X + 99$

Among them, X is the number of above-ground layers, Y is the carbon emissions per unit area, and the unit is: $kgCO_2/m^2$

	Energy Consumption Type	Carbon Emission ($kgCO_2$)	Carbon Emissions Per Unit Area ($kgCO_2/m^2$)
Carbon Emission Term	Heating	1578234.85	96
	Ventilation	7158682.12	418
	Air Conditioning	782528.23	64
	Illumination	5421576.15	322
	Equipment	51248.18	7
	Elevator	421542.17	2
Carbon Reduction Term	Photovoltai C Panels	6251484.84	371
	Building Carbon Sink	3548914.68	252
	Total	5616412.18	286

Calculation Result of Carbon Emission in Building Use Stage

Building Useage Area	Above-ground Floors	Carbon Emission Intensity ($kgCO_2/m^2$)	Building Carbon Emissions
4072	3	4.99	20.31

Energy consumption during the construction phase.

Analysis of the Energy Generation Outcomes from Renewable Sources During the Operational Phase of the Building					
Photovoltaic C Panels [$kW \cdot h/(m^2 \cdot a)$]	Green Roof	Solar Heating/Cooling [$kW \cdot h/(m^2 \cdot a)$]	Steps on the Floor to Generate Electricity	Solar Domestic Hot Water [$kW \cdot h/(m^2 \cdot a)$]	Biomass [$kW \cdot h/(m^2 \cdot a)$]	Air Source Heat Pump [$kW \cdot h/(m^2 \cdot a)$]
39.76	7.9	17.35	5.23	2.1	0	0

24 Elevation

25 Elevation Analyse

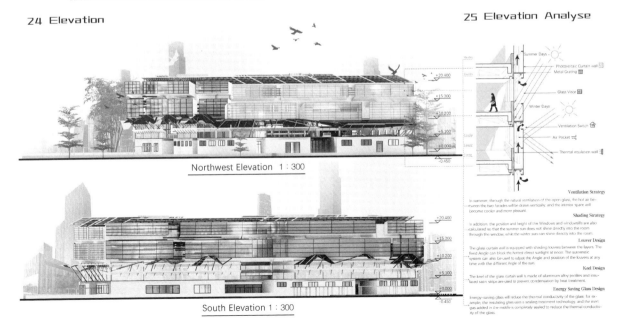

Northwest Elevation 1:300

South Elevation 1:300

Ventilation Strategy
In summer, through the natural ventilation of the open glass, the hot air between the two facades will be drawn vertically, and the interior space will become cooler and more pleasant.

Shading Strategy
In addition, the position and height of the Windows and windowsills are also calculated so that the summer sun does not shine directly into the room through the window, while the winter sun can shine directly into the room.

Louver Design
The glass curtain wall is equipped with shading louvers between the layers. The fixed Angle can block the hottest direct sunlight at noon. The automatic system can also be used to adjust the Angle and position of the louvers at any time with the different Angle of the sun.

Keel Design
The keel of the glass curtain wall is made of aluminum alloy profiles and insulated satin strips are used to prevent condensation caused by heat treatment.

Energy Saving Glass Design
Energy-saving glass will reduce the thermal conductivity of the glass: for example, the insulating glass uses a sealing treatment technology, and the inert gas added in the middle is completely sealed to reduce the thermal conductivity of the glass.

综合奖·优秀奖·零碳设计项目
Comprehensive Awards · Honorable Mention · Zero-Carbon Design Project

注册号：101438
Register Number：101438

项目名称：风之馆
Entry Title：Museum of Winds

作者：晏肇键、周言婕
Authors：Zhaojian Yan and Yanjie Zhou

作者单位：北京建筑大学
Authors from：Beijing University of Civil Engineering and Architecture

指导教师：郝石盟
Tutor：Shimeng Hao

指导教师单位：北京建筑大学
Tutor from：Beijing University of Civil Engineering and Architecture

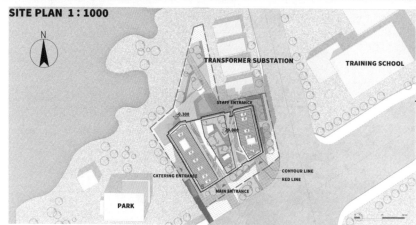

ECONOMIC INDICATORS
LAND AREA: 11447㎡
BUILDING AREA: 18200㎡
BUILDING DENSITY: 36.0%
FLOOR AREA ROTIO: 1.59

DESIGN DESCRIPTION

方案科技馆底部架空，呼应城市绿轴。主体建筑被切成三段，创造更多室外空间。南北向的视线范围被打开，在架空广场层又布置有一系列路径和场景，把室内外、上下层融为一体。

本方案采用了"空间调节"的被动式策略的若干关键设计方法，包括"适应性形体"中的空间气候梯度与形体气流组织，"交互式表皮"中的被动式气候调节腔体和分布式热虹吸自然通风系统。同时，建筑还采用了光伏板、屋顶花园、底层架空、庭院、冷巷等设计策略，保障建筑低能耗运作的同时增强建筑的热工性能。

The bottom of the science and technology museum is overhead, echoing the green axis of the city. The main building is cut into three sections to create more outdoor space. The north-south line of sight is opened, and at the overhead plaza level, a series of paths and scenes are arranged, integrating the indoor and exterior, upper and lower floors.

This proposal adopts several key design approaches of the passive strategy of "spatial conditioning", including spatial climate gradients and shape airflow organization in "adaptive shape", passive climate regulation cavity in "interactive skin", and distributed thermosyphon natural ventilation system. At the same time, the building also adopts photovoltaic panels, roof gardens, ground floor overheads, courtyards, cold alleys and other design strategies to ensure the low-enegy operation of the building and enhance the thermal performance of the building.

LOCATION ANALYSIS

The project is located at the intersection of Yushui First Road and Science and Education Avenue in the central axis and comprehensive functional area of Guangzhou Science and Education City, Zengcheng District, Guangzhou. The subtropical oceanic monsoon climate has high temperature, abundant rainfall, few frost days and sufficient sunlight. Hot and rainy, long summer without winter.

GUANGZHOU IMAGE ANALYSIS

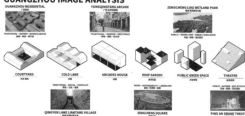

CLIMATE ANALYSIS

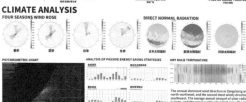

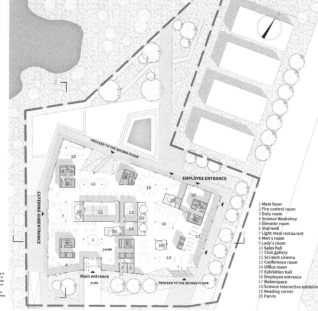

GROUND FLOOR PLAN 1:500

1 Main foyer
2 Fire control room
3 Duty room
4 Science Workshop
5 Elevator room
6 Stairwell
7 Light meal restaurant
8 Men's room
9 Lady's room
10 Sales hall
11 Civic gallery
12 Sci-tech cinema
13 Conference room
14 Office room
15 Exhibition hall
16 Employee entrance
17 Makerspace
18 Science interactive exhibition
19 Reading corner
20 Parvis

SECOND FLOOR PLAN 1:300

1 Grand step
2 Main hall
3 Ticket office
4 Leisure space
5 Elevator room
6 Stairwell
7 Maker's workshop
8 Men's room
9 Lady's room
10 Lake view deck
11 Outdoor exhibition
12 Energy-saving exhibition
13 Coffee bar
14 Parvis

SPACE CLIMATE GRADIENT [空间气候梯度]

According to the requirements of functions and environmental performance of different Spaces inside the building, the climate gradient inside the building is constructed through reasonable space configuration, so that the area with strict environmental regulation requirements is surrounded by non-strict regulation area and climate buffer zone, and the environmental regulation of the main used space is realized by the space organization itself.

The lecture hall and meeting room in the Science and Technology Museum are wrapped by other auxiliary Spaces, forming a "temperature onion". According to the different use temperature requirements of the building space, the use space is arranged in the order from the inside out, forming a certain gradient space to achieve the purpose of energy saving.

STRICT ENVIRONMENTAL CONTROL FUNCTION AREA

NON-STRICT ENVIRONMENTAL CONTROL OF PUBLIC AREAS

SPACE CLIMATE GRADIENT MODEL

CLIMATIC BUFFER ZONE

THIRD FLOOR PLAN 1:300

1 Spiral staircase atrium
2 Temporary exhibition
3 Lakeview cafe
4 Freight elevator room
5 Elevator room
6 Stairwell
7 Permanent exhibition
8 Men's room
9 Lady's room
10 Conference room
11 Activity room
12 Reception room
13 Management occupancy
14 Storeroom
15 Office
16 Outdoor terrace

Temperature curve of the coldest day in the technology cinema

Hottest day temperature curve of science and technology Cinema

Office coldest day temperature curve

Office hottest day temperature curve

The coldest day temperature curve of the foyer

The hottest day temperature curve of the foyer

A first floor spatial climate gradient

Three plane spatial climate gradients

Five plane spatial climate gradient

LOCAL RENDERING

COLD LANE GREEN VALLEY SPACE

ELEVATED OPEN SPACE

SECTION 1-1 1:300

SECTION 2-2 1:300

FOURTH FLOOR PLAN 1 : 300

1 Characteristic exhibition
3 Reading space
4 Freight elevator room
5 Elevator room
6 Stairwell
7 Atrium
8 Men's room

9 Lady's room
10 technology training room
11 Exhibit making room
12 Employee home
13 Leadership office
14 Storeroom
15 Laboratory
16 Outdoor terrace

FIFTH FLOOR PLAN 1 : 300

1 Green exhibition hall
2 Permanent exhibition
3 Popular science cinema
4 Freight elevator room
5 Elevator room
6 Stairwell
7 Curator's office
8 Men's room

9 Lady's room
10 Technology education room
11 Staff workshop
12 Back office
13 Storeroom
14 Maintenance equipment room
15 Technology reading space
16 Outdoor terrace

DISTRIBUTED THERMOSIPHON NATURAL VENTILATION SYSTEM [分布式热虹吸自然通风系统]

According to the principle of thermosiphon airflow, the air supply and exhaust tube cavities are evenly arranged inside the science and technology museum to achieve precise and efficient natural ventilation in a distributed and subtle way.

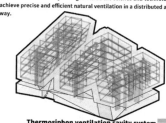

Thermosiphon ventilation cavity system
- Exhaust duct cavity
- Air duct cavity

- Air supply path
- Exhaust path
- Air supply cavity
- Exhaust cavity
- Platform air supply cavity
- Ceiling exhaust chamber

Conduit plane

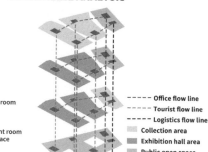

The side of the air supply shaft side diagram | Exhaust shaft side diagram

STREAMLINE ANALYSIS

- Office flow line
- Tourist flow line
- Logistics flow line
- Collection area
- Exhibition hall area
- Public open space
- Academic research platform
- Parking lot
- Equipment occupancy

STREAMLINE ANALYSIS [建筑碳排放监测智能控制系统]

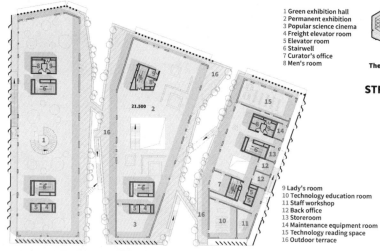

SOUTH ELEVATION 1 : 300 | **WEST ELEVATION 1 : 300**

PASSIVE CLIMATE REGULATION CHAMBER LAYER [被动式气候调节腔层]

Using the material, component and space organization design of the envelope structure in a certain thickness or depth, the design technology of interactive regulation of the wind, light and heat environment inside and outside the building.

No cavity layer midsummer indoor operating temperature

No cavity layer cooling, With the north cavity layer heating and lighting   cooling, heating and lighting

No cavity layer: annual refrigeration energy consumption of 8297kW·h, heating energy consumption of 2523kW·h
There is a north cavity layer: the annual refrigeration energy consumption is 8120kW·h, and the heating energy consumption is 2377kW·h, a total reduction of 323kW·h, accounting for 0.4%.

With the north cavity layer midsummer the summer solstice indoor operating temperature

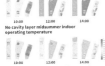

No cavity layer midsummer indoor operating temperature

No cavity layer cooling, With the east west cavity layer heating and lighting   cooling, heating and lighting

No cavity layer: annual refrigeration energy consumption of 8297kW·h, heating energy consumption of 2523kW·h
There are east and west side cavity layers: annual refrigeration energy consumption 7661kW·h, heating energy consumption 2247kW·h, a total reduction of 712kW·h, accounting for 0.6%

With the East and west cavity layer midsummer the summer solstice indoor operating temperature

No cavity layer midsummer indoor operating temperature

No cavity layer cooling, With the south cavity layer heating and lighting   cooling, heating and lighting

No cavity layer: annual refrigeration energy consumption of 8297kW·h, heating energy consumption of 2523kW·h, lighting energy consumption of 4308kW·h
There is a south cavity layer: annual refrigeration energy consumption of 7950kW·h, heating energy consumption of 2447kW·h, lighting energy consumption of 4406kW·h, a total reduction of 325kW·h, accounting for 0.4%.

With the south cavity layer midsummer the summer solstice indoor operating temperature

Light and heat balance in summer — **Winter light and heat balance** — **Winter lighting measures** — **Transitional monsoon pressure Ventilation under the mechanism** — **Ventilation under hot pressing mechanism in summer** — **Shielding east-west thermal radiation** — **Pinch ventilation carries away heat and radiates heat**

THE SOUTH SIDE AND HEAT BALANCE SHADING ADJUSTING CAVITY LAYER   **NORTH SIDE FILL LIGHT VENTILATION TO ADJUST THE CAVITY LAYER**   **EAST WEST SIDE HEAT INSULATION ADJUSTMENT CAVITY LAYER**

ANALYTICAL SIMULATION

WINTER PRESSURE CLOUD DIAGRAM

WINTER 1.5M VELOCITY VECTOR DIAGRAM

AIR FLOW DIAGRAM OF GROUND FLOOR ROOM

WIND VELOCITY VECTOR DIAGRAM

WINTER VELOCITY CLOUD DIAGRAM

SUMMER 1.5M PRESSURE VECTOR DIAGRAM

INTERIOR AIR FLOW DIAGRAM ON THE EAST SIDE

INDOOR WIND VELOCITY VECTOR DIAGRAM

GREEN ENERGY SAVING DESIGN

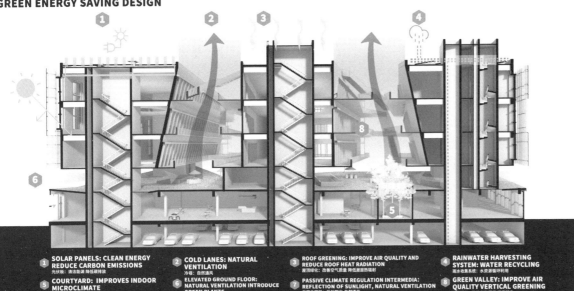

1. **SOLAR PANELS: CLEAN ENERGY REDUCE CARBON EMISSIONS** 光伏板：清洁能源 降低碳排放
2. **COLD LANES: NATURAL VENTILATION** 冷巷：自然通风
3. **ROOF GREENING: IMPROVE AIR QUALITY AND REDUCE ROOF HEAT RADIATION** 屋顶绿化：改善空气质量 降低屋顶热辐射
4. **RAINWATER HARVESTING SYSTEM: WATER RECYCLING** 雨水收集系统：水资源循环利用
5. **COURTYARD: IMPROVES INDOOR MICROCLIMATE** 天井庭院：改善室内微气候
6. **ELEVATED GROUND FLOOR: NATURAL VENTILATION INTRODUCE GREEN PLANTS** 底层架空：自然通风 引入绿植
7. **PASSIVE CLIMATE REGULATION INTERMEDIA: REFLECTION OF SUNLIGHT, NATURAL VENTILATION** 东西向腔层：反射阳光 自然通风
8. **GREEN VALLEY: IMPROVE AIR QUALITY VERTICAL GREENING** 生态绿谷：改善空气质量 垂直绿化

综合奖・优秀奖・零碳设计项目
Comprehensive Awards · Honorable Mention · Zero-Carbon Design Project

注册号：101457
Register Number：101457

项目名称：接・融
Entry Title：Connection, Fusion

作者：徐一、艾力克尔・艾百都拉、陈浩
Authors：Yi Xu, Ailiker. Aibaidula and Hao Chen

作者单位：浙江理工大学
Authors from：Zhejiang Sci-Tech University

指导教师：邓小军、文强、卜昕翔
Tutors：Xiaojun Deng, Qiang Wen and Xinxiang Bu

指导教师单位：浙江理工大学、同济大学建筑设计研究院（集团）有限公司
Tutors from：Zhejiang Sci-Tech University, and Tongji Architectural Design (Group) Co., Ltd.

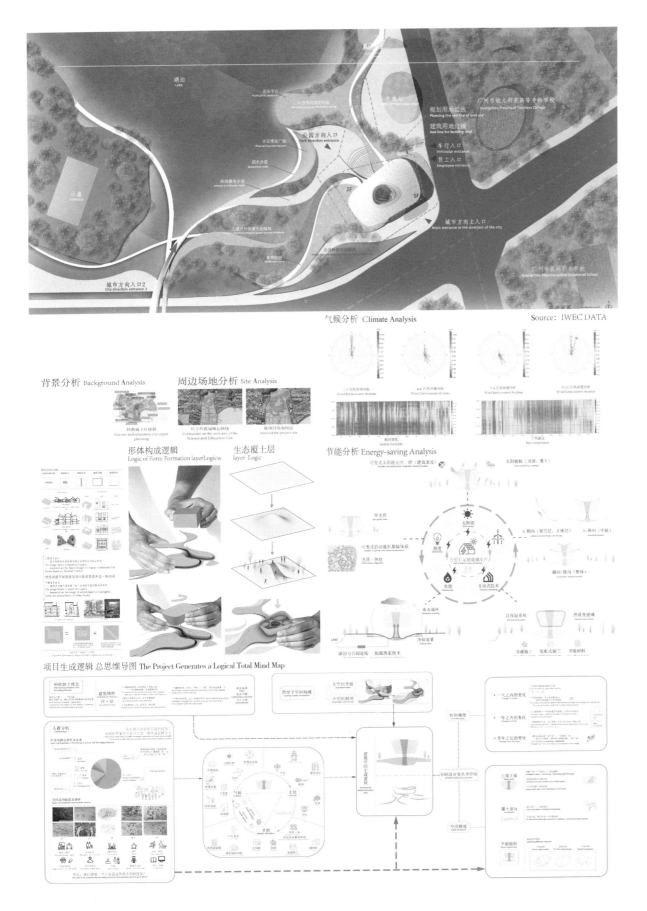

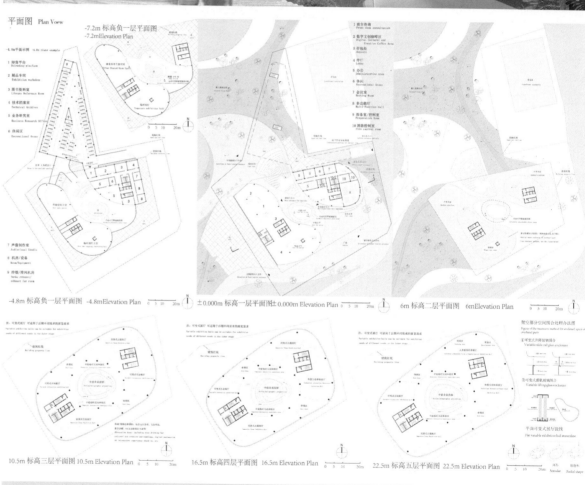

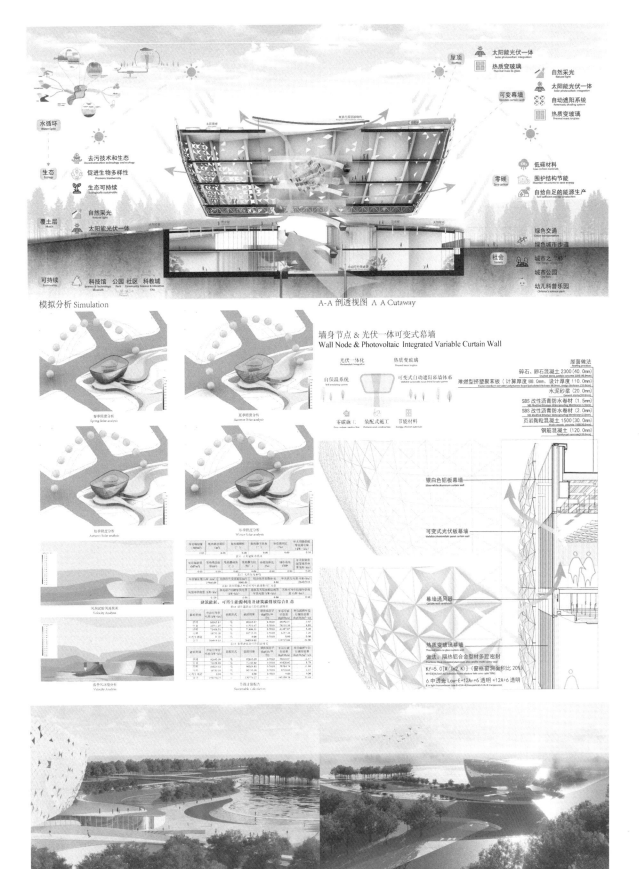

综合奖·优秀奖·零碳设计项目
Comprehensive Awards · Honorable Mention · Zero-Carbon Design Project

注册号：101463
Register Number：101463

项目名称：多级绿洲
Entry Title：Oasis

作者：钟玉蒜、程晓华、桑瑞雷
Authors：Yulin Zhong, Xiaohua Cheng and Ruilei Sang

作者单位：北京交通大学
Authors from：Beijing Jiaotong University

指导教师：张文、周艺南
Tutors：Wen Zhang and Yinan Zhou

指导教师单位：北京交通大学
Tutors from：Beijing Jiaotong University

This design aims to continue the rich spatial experience of the traditional public buildings in Guangzhou, and together with the northern park, form a recreational center for citizens. And through the hierarchical design of vertical space and interface, as well as a rich planting system, a multi-level energy utilization chain is formed in the vertical direction. In order, photovoltaic modules - Urban agriculture planting system - subtropical Broad-leaved tree - subtropical shrubs - semi indoor active space - rooms with high requirements for shading and heat insulation. In the horizontal direction, active and passive technologies are used to form a climate buffer layer, providing comfortable green and rich activity space for people moving indoors and in gray space.

本设计希望延续广州地区公共建筑中丰富的空间体验，与北侧公园共同形成市民的休憩中心。并且通过竖向空间和界面的分级设计与丰富的种植体系，在垂直方向上形成了能量的多级利用链条。依次是光伏组件-都市农业种植体系-亚热带阔叶树-亚热带灌木-半室内积极活动空间-遮光隔热要求较高的房间，水平方向上利用主被动技术形成气候缓冲层，为室内和灰空间内活动的人们提供舒适、绿色、丰富的活动空间。

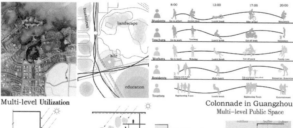

Colonnades have many applications in both Chinese and foreign architecture. They not only provide shade and rain protection, but also create a transitional space between indoor and outdoor spaces. There are many examples of colonnade buildings in Guangzhou, which are rainy and hot, forming a climate buffer zone in the horizontal direction.

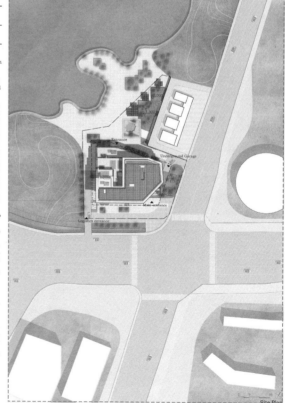

The construction site is located at a road intersection, backed by a park.

Confirm that the main entrance is located on the north side, within the building volume, leaving space inside to connect the park and the city.

About five floors of the building are calculated according to the Floor area ratio.

Place it in a ventilated atrium and adjust the seven story overhead and number of floors.

The northward retreat connects with the park and increases the possibility of multi-level utilization of building energy, while the southward self

Adjust the retreat according to functional requirements to enrich the gray space.

Incorporate vertical traffic and outer circular traffic to enrich the landscape while shading.

Placing on planting platforms and yards.

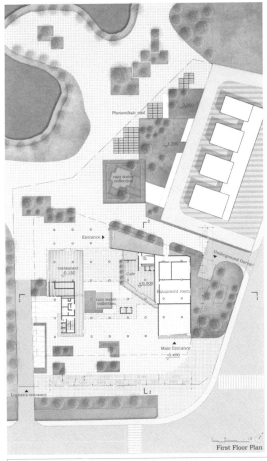

First Floor Plan

Second Floor Plan

Fifth Floor Plan

Third Floor Plan

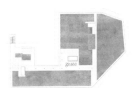
Sixth Floor Plan

Fourth Floor Plan

Roof Plan

Public Space

Energy Grading

| Urban context *disparate* | First floor commercial *extroverted* | Public activity entrance *multifunctional* | Outer circular traffic *continuous* | Large outdoor space *abundant* | Set-back model *"Hanging Garden"* | Labor space *practical* |

In order to fully utilize the solar energy of the site, we have formed a multi-level energy utilization chain in the vertical direction based on the demand for light and the utilization rate of solar energy for different activities and functions. PV modules - Urban agriculture planting system - subtropical Broad-leaved tree - subtropical shrubs - semi indoor active space - rooms with high requirements for shading and heat insulation. Solar energy is utilized and filtered layer by layer within the system, and when it reaches the main activity space of humans, both the energy and thermal energy have been reduced to be suitable for human activities.

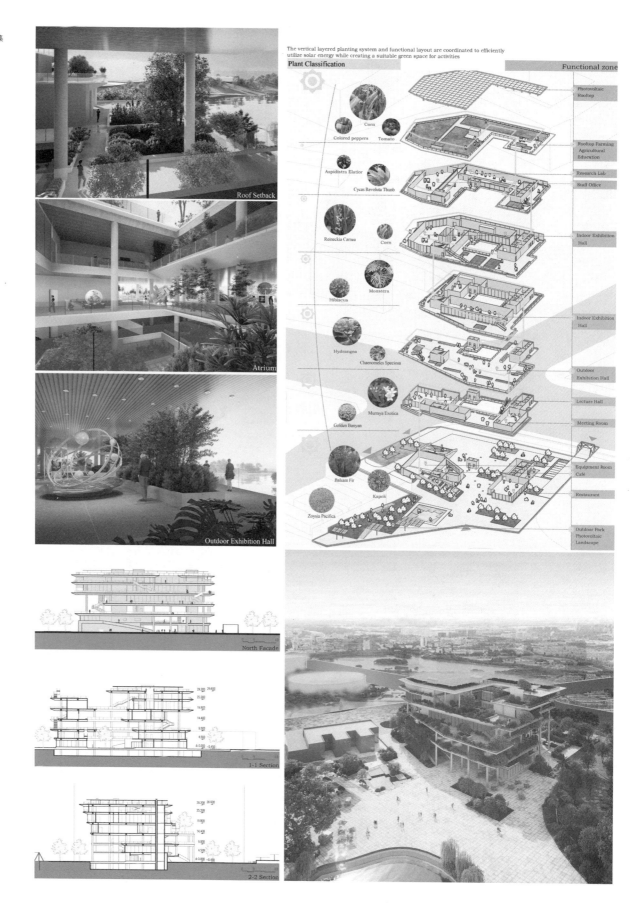

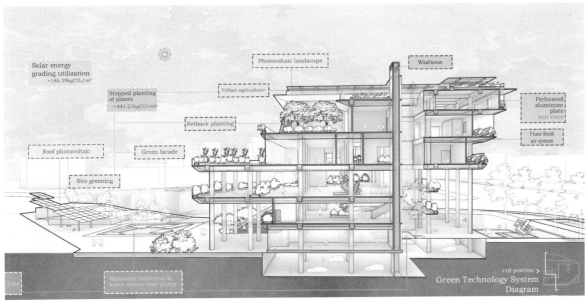

Green Technology System Diagram

Climatic Conditions

Wind Simulation

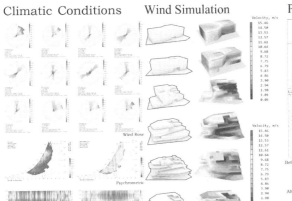

Proactive Green Technology & Energy Consumption

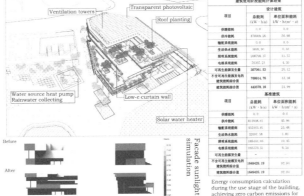

Energy consumption calculation during the use stage of the building, achieving zero carbon emissions for the entire year of the building

Details

Building Construction

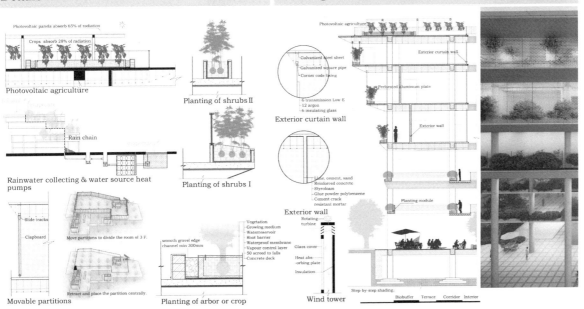

综合奖・优秀奖・零碳设计项目
Comprehensive Awards · Honorable Mention · Zero-Carbon Design Project

注册号：101584
Register Number：101584

项目名称：裂缝解码
Entry Title：GENO ART X

作者：陈蓉
Author：Rong Chen

作者单位：湖北工业大学
Author from：Hubei University of Technology

指导教师：李竞一
Tutor：Jingyi Li

指导教师单位：华中科技大学
Tutor from：Huazhong University of Science and Technology

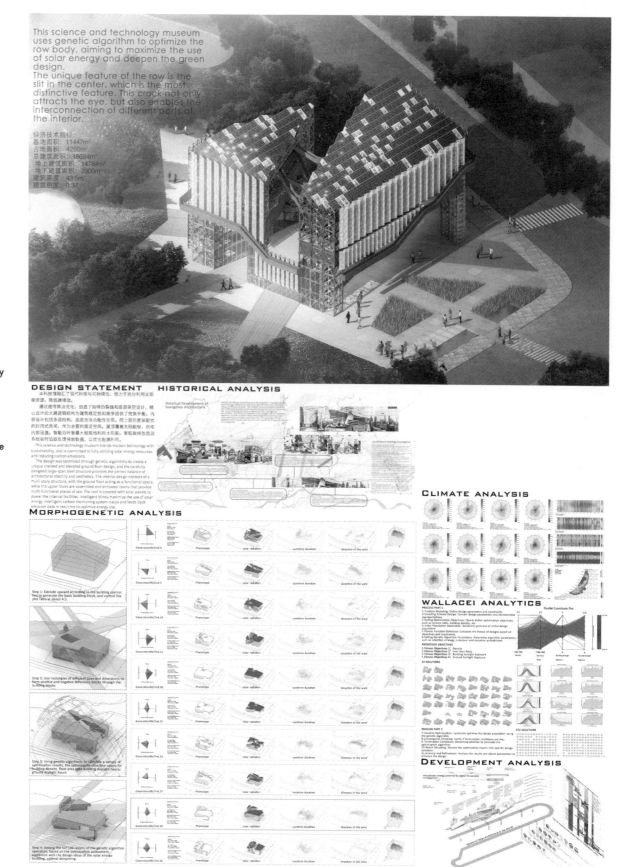

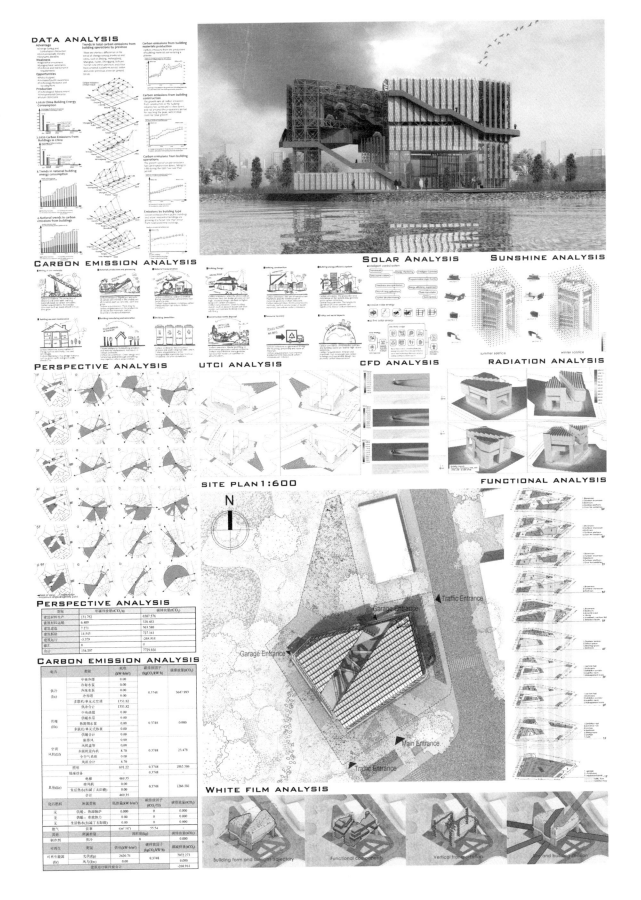

STRUCTURAL ANALYSIS

Scene Analysis

Ground Floor Overhead | Split Corridor | Cracked Stairs | Mezzanine Gallery | Outdoor Platform | Upper Exhibition Hall

First Floor Plan 1:350

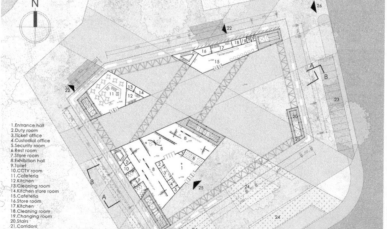

1. Entrance hall
2. Duty room
3. Ticket office
4. Custodial office
5. Security room
6. Rest room
7. Store room
8. Exhibition hall
9. Toilet
10. CCTV room
11. Cafeteria
12. Kitchen
13. Cleaning room
14. Kitchen store room
15. Cafeteria
16. Store room
17. Kitchen
18. Cleaning room
19. Changing room
20. Stairs
21. Corridors
22. Garage entrance
23. Ground floor parking
24. Flower bed
25. Exhibition entrance
26. Site entrance

Five-story floor plan

Nine-storey floor plan

North Elevation Plan 1:350   West Elevation Plan 1:350

HANDMADE MODEL

NODAL ANALYSIS

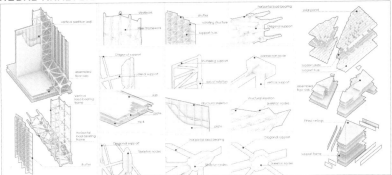

B-B SECTION VIEW 1:400

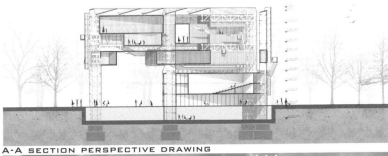

A-A SECTION PERSPECTIVE DRAWING

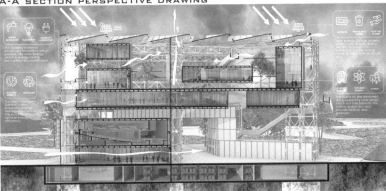

综合奖・优秀奖・零碳设计项目
Comprehensive Awards · Honorable Mention · Zero-Carbon Design Project

注册号：101611
Register Number：101611

项目名称：光之舞・荔园
Entry Title：Dance of Lights, Li Garden

作者：叶碧珊、梁振彪、吴树祺、袁鑫桐、郑泽科、蔡锶琦、刘依婷、杨文锦、蒋宇扬

Authors：Bishan Ye, Zhenbiao Liang, Shuqi Wu, Xintong Yuan, Zeke Zheng, Siqi Cai, Yiting Liu, Wenjin Yang and Yuyang Jiang

作者单位：广东工业大学
Authors from：Guangdong University of Technology

指导教师：吉慧、林瀚坤、王树希、董涧清、邓寄豫

Tutors：Hui Ji, Hankun Lin, Shuxi Wang, Jianqing Dong and Jiyu Deng

指导教师单位：广东工业大学、珠海中建兴业绿色建筑设计研究院有限公司、广州市图鉴城市规划勘测设计有限公司

Tutors from：Guangdong University of Technology, Zhongjian Xingye Green Architecture Design and Research Institute Limited of Zhuhai City and Guangzhou Tujian Urban Planning Survey and Design Co., Ltd.

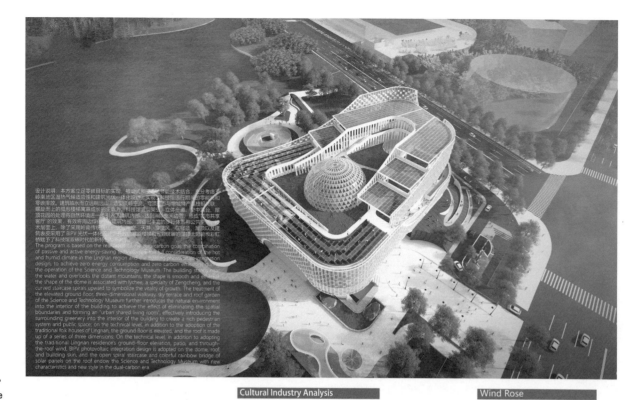

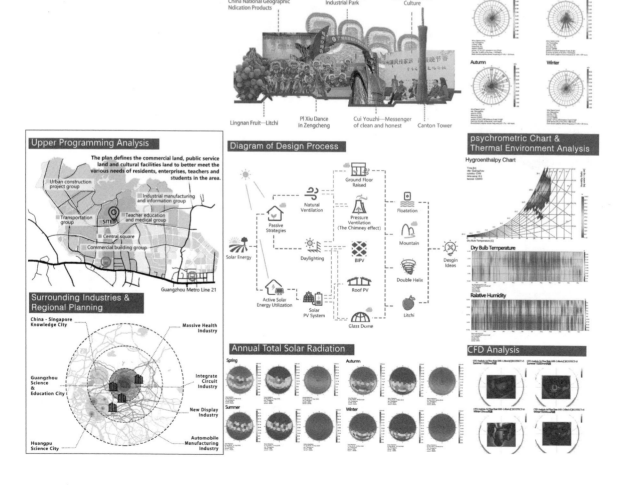

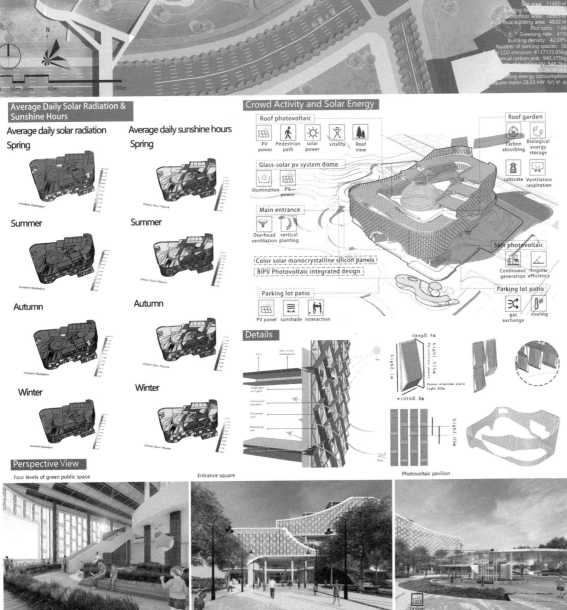

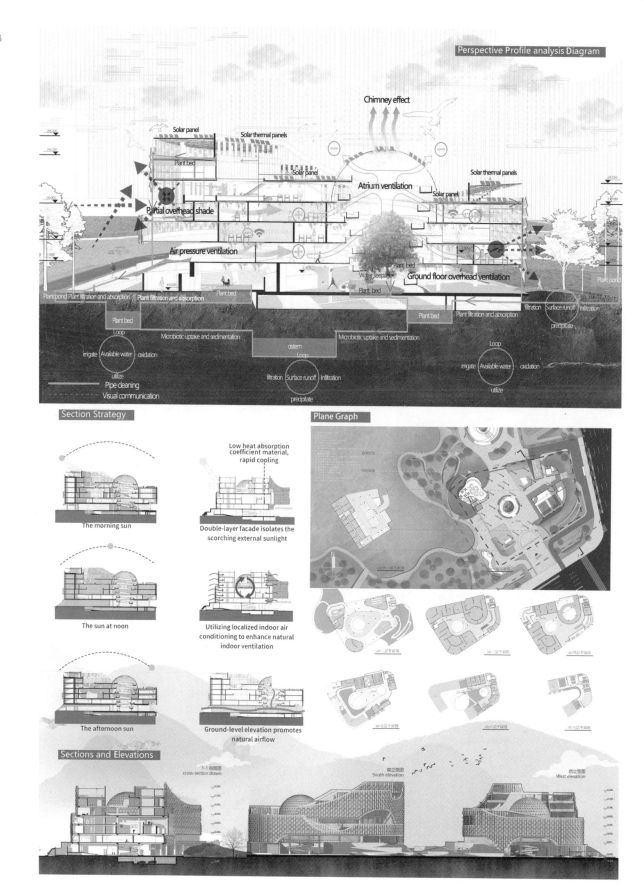

Streamline analysis

Ananlysis of Solar Energy

Building Carbon Emissions Calculation

综合奖·优秀奖·零碳设计项目
Comprehensive Awards · Honorable Mention · Zero-Carbon Design Project

注册号：101688
Register Number：101688

项目名称：转译竹筒屋
Entry Title：Bamboo Tube Transforming

作者：巫启隽、李泽清、王祎铭
Authors：Qijun Wu, Zeqing Li and Yiming Wang

作者单位：厦门大学
Authors from：Xiamen University

指导教师：韩洁
Tutor：Jie Han

指导教师单位：厦门大学
Tutor from：Xiamen University

广州科教城零碳文化科技馆设计
Design of Zero Carbon Culture and Technology Museum in Guangzhou

Location Analysis Map

The project is located in the central axis and comprehensive functional area of Guangzhou Science and Education City, Zengcheng District, Guangzhou.

The project is located in the subtropical maritime monsoon climate, with high temperature, abundant rainfall, few frost days and sufficient sunlight. The annual average temperature is 21.6℃, the extreme high temperature is 38.2℃, the extreme low temperature is -1.9℃, and the average annual rainfall is 2039.5 mm. The general climate is characterized by hot, rainy, long summer without winter.

Temperature and Humidity Conditions and Technical Treatment

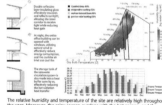

The relative humidity and temperature of the site are relatively high throughout the year. Moreover, the rainy season overlaps with the hot season, and the water brought by rainfall is quickly evaporated by the high temperature and increases humidity, resulting in frequent occurrence of humid and stuffy conditions in summer. We have made corresponding strategies to address these issues on the site, such as using roof panels and double glazed units to reduce insulation, using natural ventilation for cooling, and collecting rainwater for recycling as a coolant.

CFD Model Ventilation Simulation

We learned the cooling effect of bamboo tube house, and conducted wind velocity fluid experiment with the software. We found that the ventilation and cooling efficiency of bamboo tube house was very good, so we extracted its principle and applied it to our profile layout.

Greening Strategy Analysis Chart

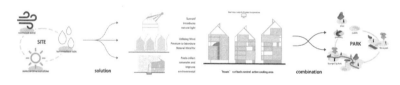

DESIGN CONCEPT

1.SITE
The climate in Guangzhou is hot, and local buildings have formed unique styles to adapt to the climate

2.TRANSLATION
We have translated the spatial layout of traditional bamboo tube houses

3.SYSTEM
Separate the exhibition area from the pedestrian system and actively regulate it separately

4.ENERGY CONSERVATION
Determine the orientation of the shape and the position of the solar panel based on the simulation of the light, heat, and wind environment

Master Plan 1:800

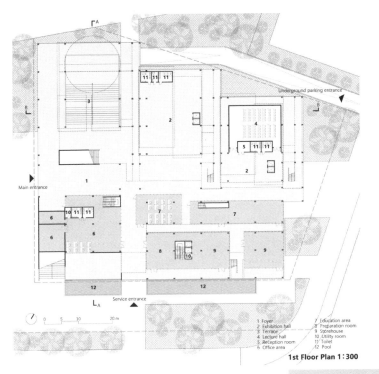

1 Foyer
2 Exhibition hall
3 Terrace
4 Lecture hall
5 Reception room
6 Office area
7 Education area
8 Preparation room
9 Storehouse
10 Utility room
11 Toilet
12 Pool

1st Floor Plan 1:300

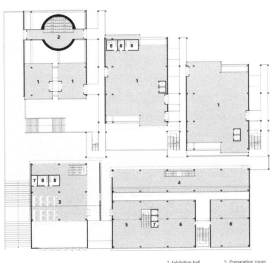

1 Exhibition hall
2 Spherical theatre
3 Office area
4 Education area
5 Preparation room
6 Storehouse
7 Utility room
8 Toilet

2nd Floor Plan 1:300

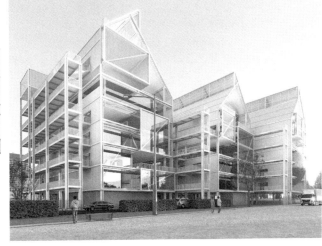

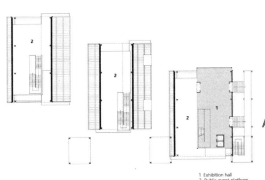

1 Exhibition hall
2 Public event platform

7th Floor Plan 1:300

LIGHTING COMPARISON

 6:00 am
 10:00 am
 2:00 pm
 6:00 pm

Designed orientation | #1 rotated | #2 rotated | #3 rotated

AIRFLOW SIMULATION

 0.5m high on the 2nd floor
 1.5m high on the 2nd floor
 2.5m high on the 2nd floor
 3.5m high on the 2nd floor

DESIGN DEVELOPMENT

1 FUNCTION
The site has been divided based on functional requirements and the wind environment of the site

2 TRANSLATION
Translate the bamboo tube house, extract the passive ventilation form from it, and place it in the building to achieve better ventilation

3 ARRANGEMENT
We adopt an active adjustment method for the exhibition hall area, separating the pedestrian system, which is different from traditional science and technology museums and reduces energy consumption

4 PUBLIC
Adding a roof to the exhibition area naturally shapes the outdoor exhibition area and public space. The roof enhances passive ventilation and heat dissipation, while also making better use of solar energy

5 SOLAR POWAE
Arrange solar panels based on simulated sunlight to maximize the utilization of solar energy and achieve zero carbon emissions

EXPLODED AXONOMETRIC & MATERIALS

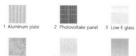

1 Aluminum plate 2 Photovoltalte panel 3 Low-E glass
4 Concrete 5 Polycarbonate 6 Steel

South Elevation 1 : 600   West Elevation 1 : 600

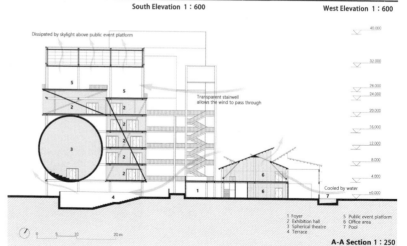

1 Foyer 5 Public event platform
2 Exhibition hall 6 Office area
3 Spherical theatre 7 Pool
4 Terrace

A-A Section 1 : 250

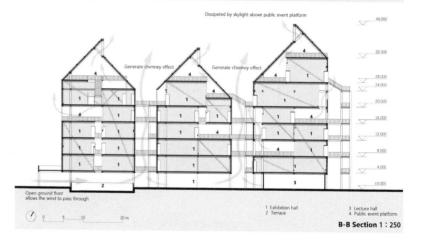

1 Exhibition hall 3 Lecture hall
2 Terrace 4 Public event platform

B-B Section 1 : 250

2023 台达杯国际太阳能建筑设计竞赛获奖作品集

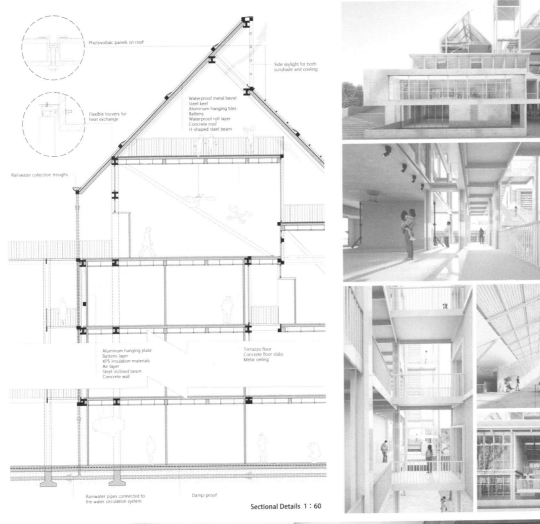

Sectional Details 1:60

085

综合奖·优秀奖·零碳设计项目
Comprehensive Awards · Honorable Mention · Zero-Carbon Design Project

注册号：101710
Register Number：101710

项目名称：管叠·风起
Entry Title：Ducts Play, Winds Arise

作者：黄泽宇、程雅雯、胡霞胜、缪珂
Authors：Zeyu Huang, Yawen Cheng, Xiasheng Hu and Ke Miu

作者单位：南京工业大学
Authors from：Nanjing Tech University

指导教师：舒欣
Tutor：Xin Shu

指导教师单位：南京工业大学
Tutor from：Nanjing Tech University

TECHNICAL ECONOMIC INDICATOR
- Land Area:
- Overall floorage: 4352 ㎡
- Building Area: 17744 ㎡
- Building density: 1.22
- Building Height: 27.8 m

DESIGN GENERATION

Climate characteristics analysis
1. The climate is hot in summer and mild in winter
2. Monsoonal climate
3. Occasional typhoon impact during autumn
4. Abundant sunlight, Adequate precipitation

Courtyard house / Bamboo hut — Conceptual translation
Extended roof overhang blocks sunlight / Narrow alley aids ventilation — Cold lane / Functional space

Shaping the building with courtyard form to create an inner courtyard

Translating "bamboo hut" into "a massing + cold lane pipe system", stacking pipes to form a partial inner courtyard at the overlaps for ventilation

DESIGN DESCRIPTION

我们的目的是以"风管"为主题，打造一个绿色开放的，具有岭南地方特色的科技馆。

我们从围屋、竹筒屋等传统民居转译出风管概念，根据"气候赋型"理论在建筑中加入横纵向的风管，横纵风管分别"引风"与"促风"，横纵风管交错形成大小中庭等多样的空间，管上空间与管内空间丰富了人流的动线。用管内和管外流线的组织，带给参观者时空交错般的空间体验。

建筑立面和屋顶铺设太阳能光伏板，天窗玻璃选用光伏玻璃，建筑内加入了智能化控制系统，场地运用了海绵城市、水处理等主、被动太阳能技术，让游客在建筑内达到舒适度要求的同时，也大大减少了建筑能耗。

Our objective is to establish a technologically advanced and environmentally sustainable science museum that showcases the unique culture of the Lingnan region.

We translated air ducts from traditional houses such as enclosed houses and bamboo hut. By extrapolating suitable dimensions for air ducts based on the principles of 'climate typology', we have integrated both horizontal and vertical ventilation ducts into our building design. These ducts serve the dual functions of 'inflow' and 'promoting airflow' respectively. Their intersection creates varying courtyard sizes and others diverse range of space, enhancing the circulation of people. And the organization of both interior and exterior ducts results in a spatial experience that intertwines time and space, offering visitors a dynamic and immersive environment.

The building's facades and roof are equipped with solar photovoltaic panels, while the skylights feature photovoltaic glass. Additionally, the structure incorporates an intelligent control system. The site design incorporates both active and passive technologies such as sponge city and water treatment. These measures not only ensure visitor comfort within the building but also significantly reduce its energy consumption.

CLIMATE ANALYSIS

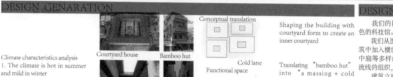

Annual temperature / Solar radiation / Psychrometric chart
Relative humidity / Daily temperature / Direct Sun Hours / Solar path

OPTIMAL DUCTS SHAPE GENERATION

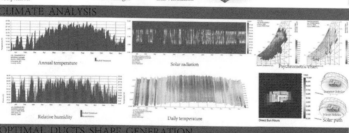

1. Advantage: Creates airflow circulation within the spatial structure (hexagram).
2. Poor horizontal ventilation transfer effect.
3. Poor vertical wind transfer effect.
4. Forms a loop but with lower efficiency than ours.

CONCEPTUAL DESIGN

1. Combine air ducts into an L-shaped formation
2. Make the formation effective to ventilation
3. Fit ducts according to functional and circulation requirements
4. Cutting the building along the landscape axis

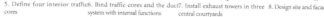

5. Define four interior traffic cores
6. Bind traffic cores and the duct system with internal functions
7. Install exhaust towers in three central courtyards
8. Design site and facades

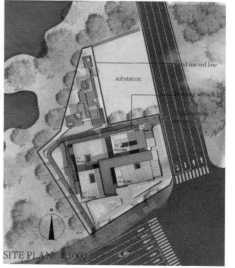

SITE PLAN 1:1000

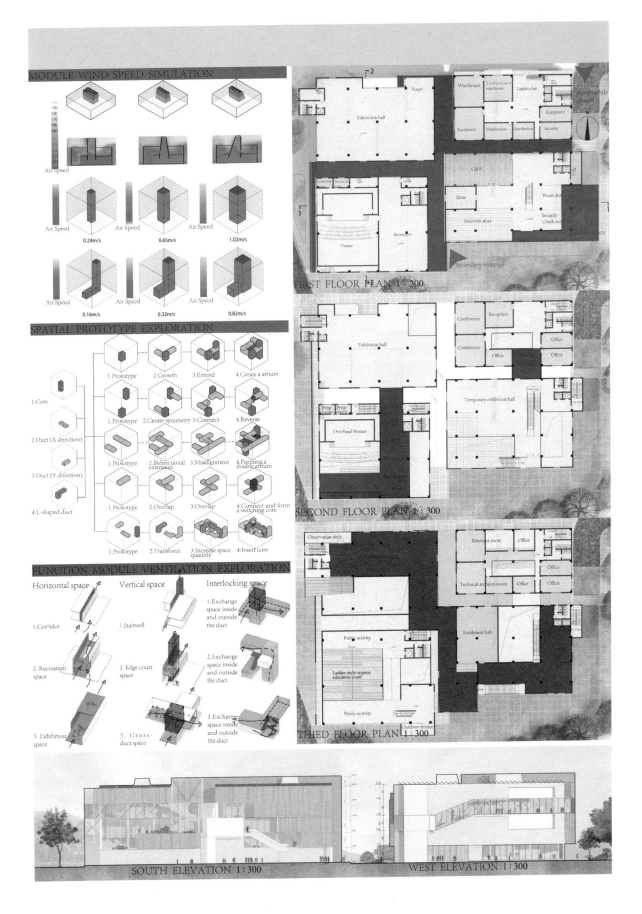

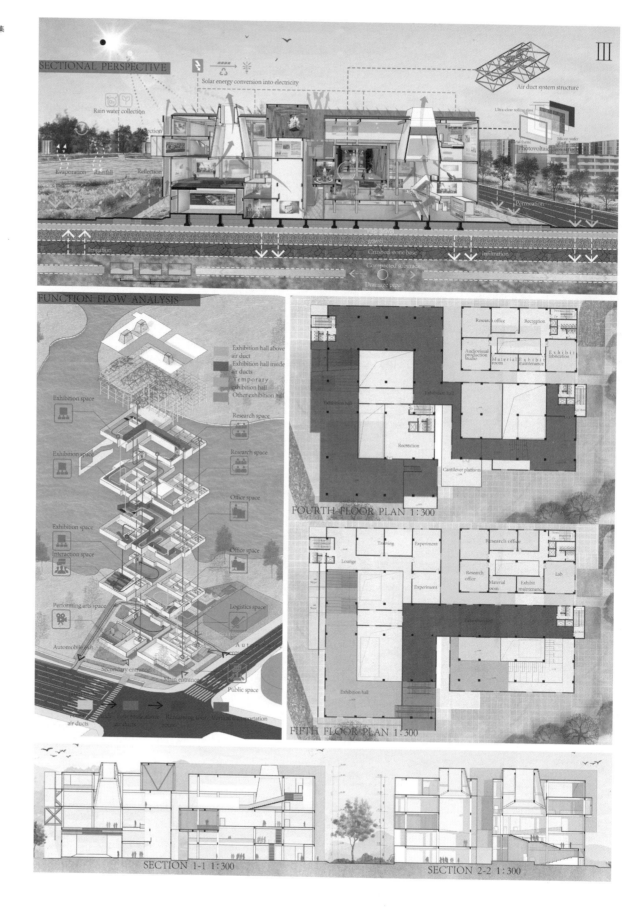

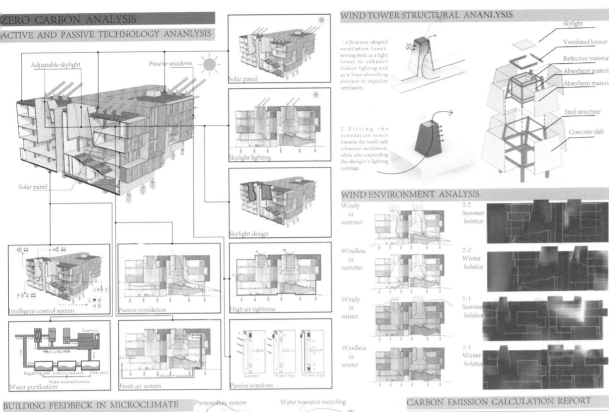
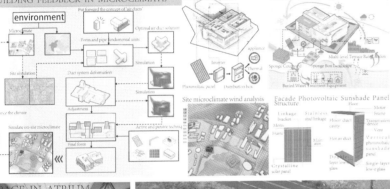

综合奖・优秀奖・零碳设计项目
Comprehensive Awards・Honorable Mention・Zero-Carbon Design Project

注册号：101811
Register Number：101811

项目名称：向阳而生
Entry Title：Growing towards the Sun

作者：宋郭睿、董春朝、王艺洋、张吉森、李文强、赵鑫慧、张梦婧、张一鸣、贺富年

Authors：Guorui Song, Chunchao Dong, Yiyang Wang, Jisen Zhang, Wenqiang Li, Xinhui Zhao, Mengjing Zhang, Yiming Zhang and Funian He

作者单位：西北工业大学、中铁十二局数字土木分公司

Authors from：Northwestern Polytechnical University and Digital & Civil Engineering Branch, No.12 Bureau, China Railway Engineering Corp

指导教师：刘煜、宋戈、曹建
Tutors：Yu Liu, Ge Song and Jian Cao

指导教师单位：西北工业大学
Tutors from：Northwestern Polytechnical University

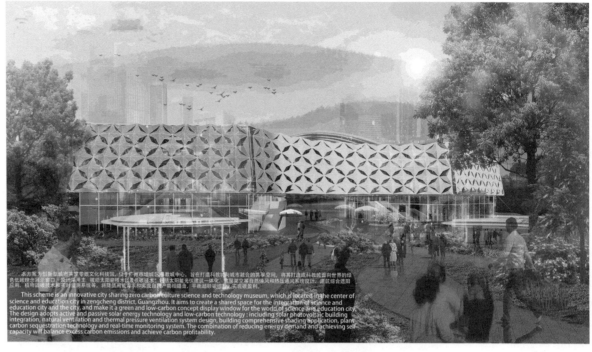

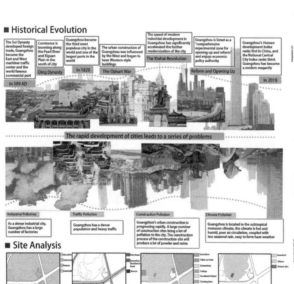

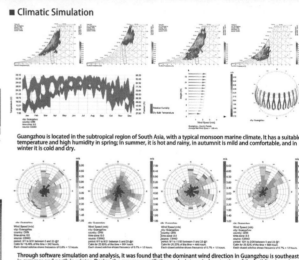

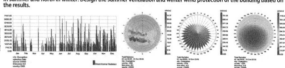

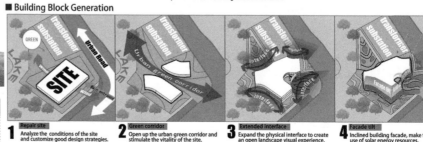

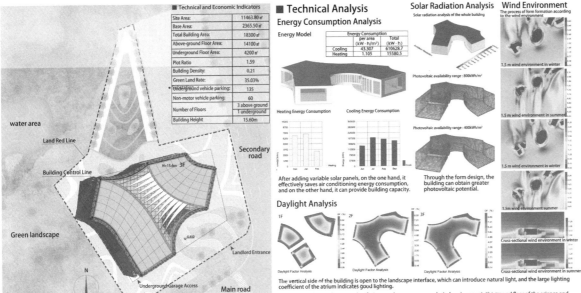
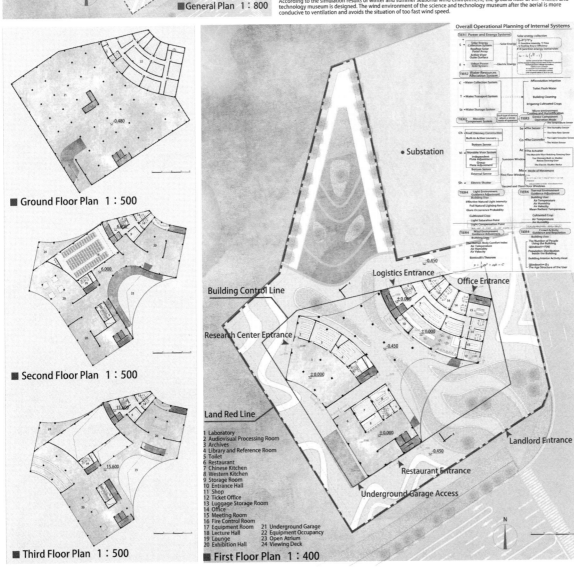

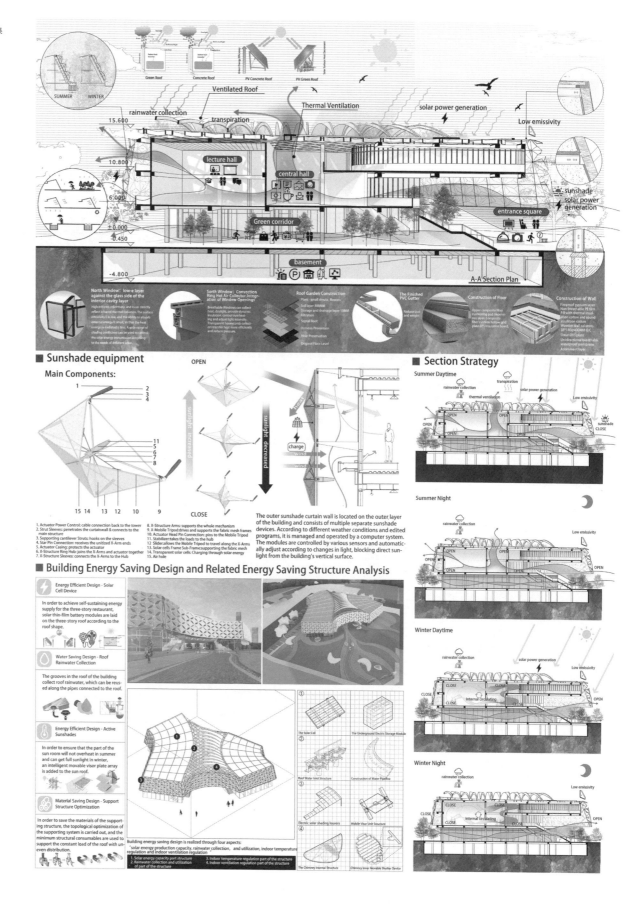

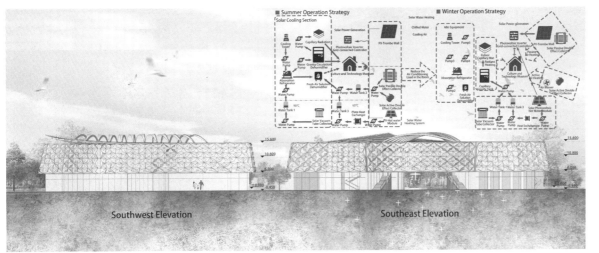

Southwest Elevation | Southeast Elevation

■ Carbon Emission Logic Diagram

■ Annual Carbon Emissions Accounting

Category	Subkey	Area(㎡)	Carbon emission (kg/a)	Subtotal (kg/a)	Total (kg/a)
Energy demand	Air conditioner	/	907575.5	2261291.5	-8293.6
	Illumination	/	686019.5		
	Elevator operation	/			
	Cooking	/	667696.5		
	Information equipment	/			
	Others	/			
Renewable energy	Roof BIPV	2520.1	-1265400	-2257400	Purchase green electricity through carbon trading to balance excess carbon emissions
	Facade BIPV	1975.6	-992000		
Carbon sink	Arbor	456.1	-4282.8	-12185.1	
	Bush	513.4	-1293.8		
	Herbal	703.78	-6608.5		

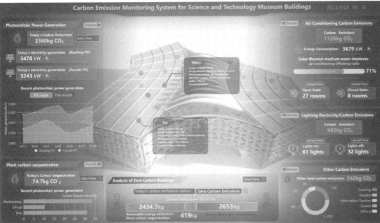

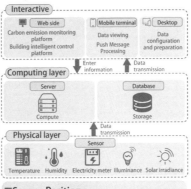

■ Digital Platform Framework Diagram

■ Sensor Position

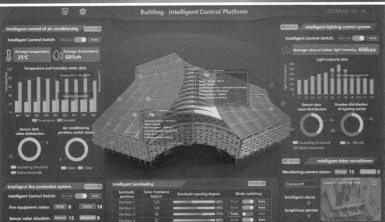

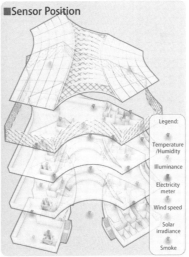

Legend:
- Temperature/Humidity
- Illuminance
- Electricity meter
- Wind speed
- Solar irradiance
- Smoke

综合奖·优秀奖·零碳设计项目
Comprehensive Awards · Honorable Mention · Zero-Carbon Design Project

注册号：101926
Register Number：101926

项目名称：穿竹
Entry Title：Through the Bamboo

作者：谢丛朵、叶睿文、郑丹彤、唐国言
Authors：Congduo Xie, Ruiwen Ye, Dantong Zheng and Guoyan Tang

作者单位：苏州大学
Authors from：Soochow University

指导教师：孙磊磊、王彪、韩冬辰
Tutors：Leilei Sun, Biao Wang and Dongchen Han

指导教师单位：苏州大学
Tutors from：Soochow University

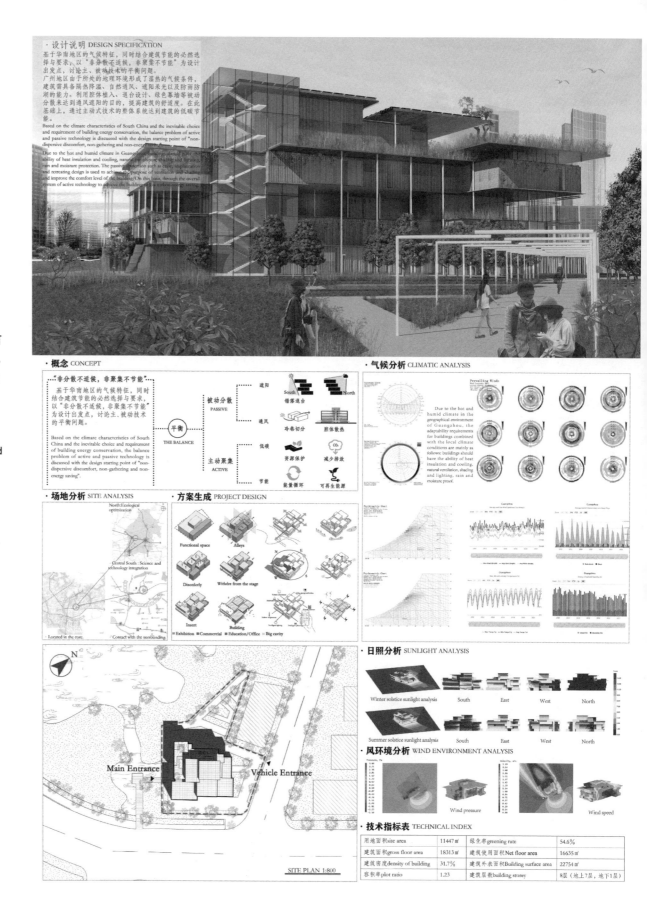

· 爆炸图 EXPLODEDE VIEW

· 植被分析 VEGETATIONAL ANALYSIS

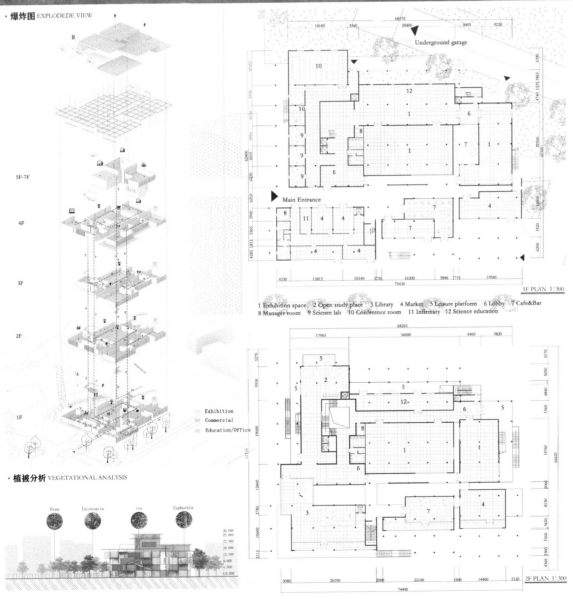

1 Exhibition space 2 Open study place 3 Library 4 Market 5 Leisure platform 6 Lobby 7 Cafe&Bar
8 Manager room 9 Science lab 10 Conference room 11 Infirmary 12 Science education

1F PLAN 1:300

2F PLAN 1:300

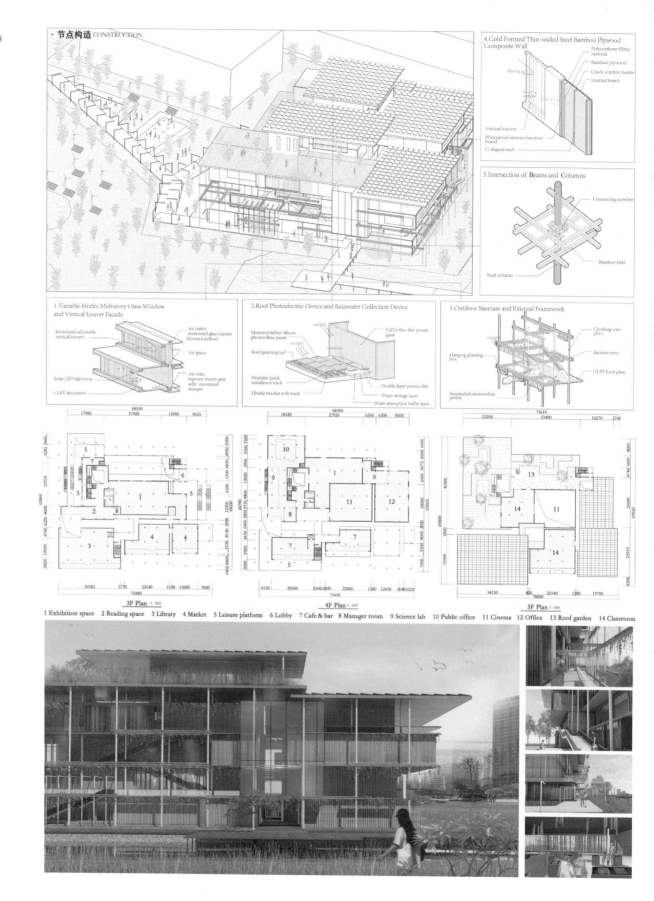

· 全生命周期分析 FULL LIFE CYCLE ANALYSIS

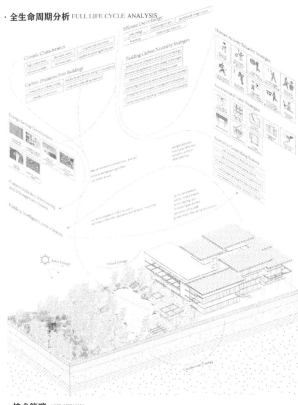

· 技术策略 STRATEGY

· 碳排放与能耗计算 CALCULATIONS

建筑使用阶段碳排放计算结果
Calculation results of carbon emissions during service stage

Project		Amount of carbon emissions (kgCO₂)	Amount of carbon emissions per unit area (kgCO₂/m²)
Carbon emission	HVAC	11602378.72	633.54
	hot water	4609967.85	251.72
	lighting	296944.75	16.21
	lift system	854163.64	46.64
Carbon reduction	renewable energy	5912241.48	322.83
	carbon sink	1895848.50	103.52
Summation		9555364.98	521.76

使用阶段可再生能源产能量计算结果
Calculation results of renewable energy production during servive stage

Photovoltaics [kW·h/(m²·a)]	31.85
Solar power supply Warm/Cold [kW·h/(m²·a)]	0.37
Solar energy generation Active hot water [kW·h/(m²·a)]	0.41
Ground Source Heat Pump [kW·h/(m²·a)]	0.36

建筑能耗技术指标计算结果
Calculation results of building energy consumption technical indicators

Project	Comprehensive energy saving rate (%)	Utilization of renewable energy (%)	Energy efficiency rate of building body (%)	Heating cooling and lighting Average energy consumption index (kW·h/m²·a)	Reduction in carbon emission intensity (kg CO₂/m²·a)
Figures	50.39	33.07	26.98	19.76	13.90

After software simulation and calculation, the project technology we designed can meet the requirements of "Technical standard for nearly zero energy buildings". Therefore, we can achieve ultra-low energy consumption buildings.

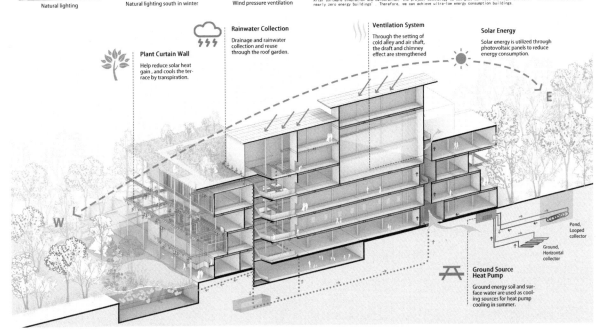

综合奖·优秀奖·零碳设计项目
Comprehensive Awards · Honorable Mention · Zero-Carbon Design Project

注册号：101972
Register Number：101972

项目名称：天上星河转
Entry Title：Turning Galaxy

作者：黄泽坤、孙怡恒、吴茗、方心琪、罗翔
Authors：Zekun Huang, Yiheng Sun, Ming Wu, Xinqi Fang and Xiang Luo

作者单位：河南大学
Authors from：Henan University

指导教师：康永基、宗慧宁
Tutors：Yongji Kang and Huining Zong

指导教师单位：河南大学
Tutors from：Henan University

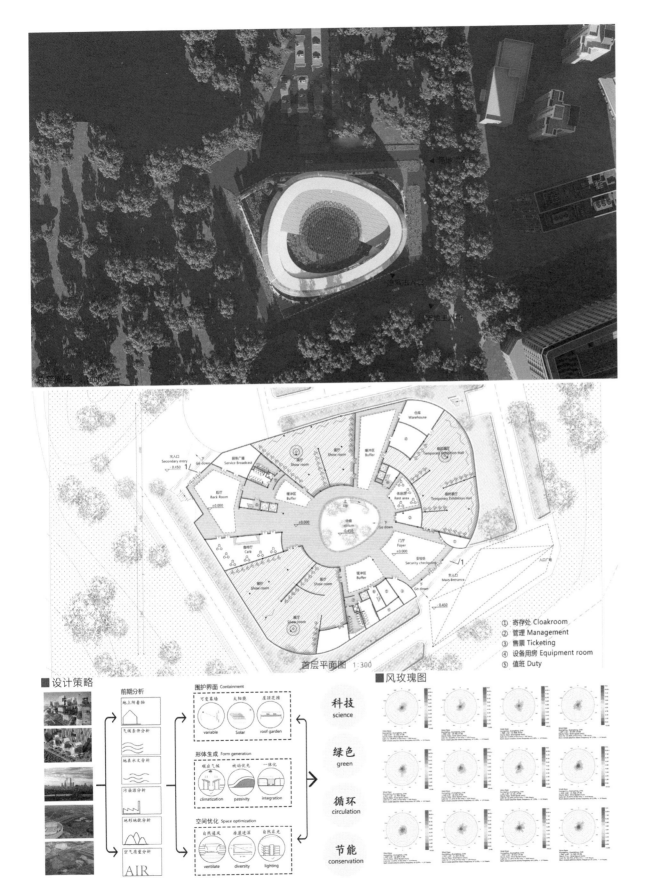

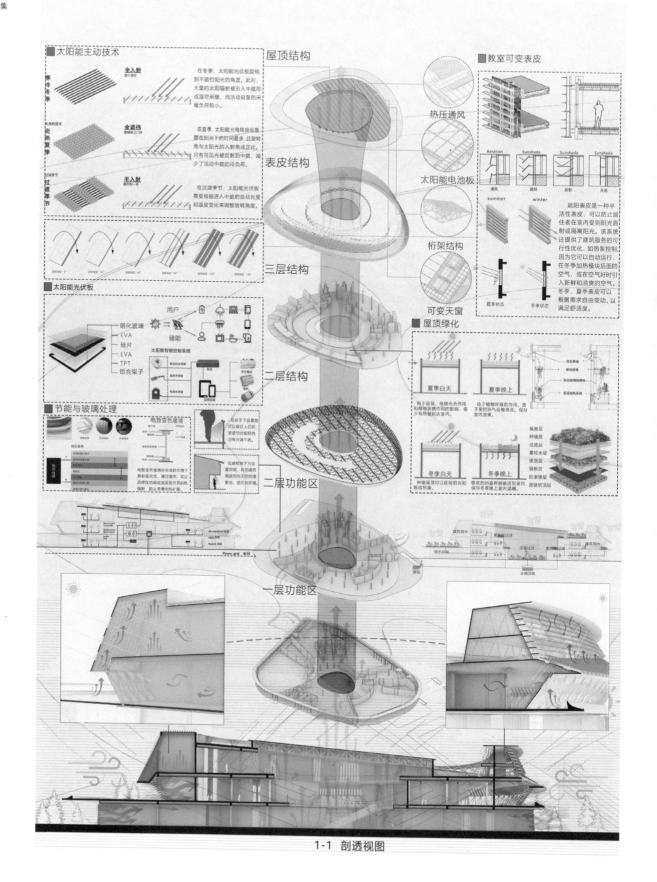

1-1 剖透视图

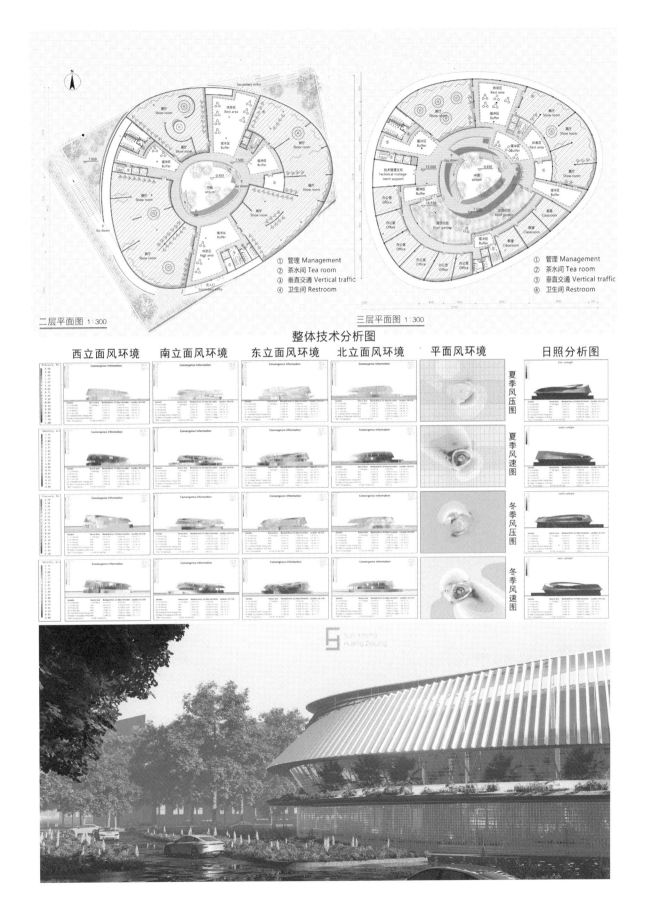

综合奖·优秀奖·零碳设计项目
Comprehensive Awards · Honorable Mention · Zero-Carbon Design Project

注册号：102021
Register Number：102021

项目名称：风声不息
Entry Title：The Wind Never Dies Down

作者：刘阳、刘菊影、张金飞、刘晓琪、辛瑞铭
Authors：Yang Liu, Juying Liu, Jinfei Zhang, Xiaoqi Liu and Ruiming Xin

作者单位：河南大学
Authors from：Henan University

指导教师：康永基、宗慧宁
Tutors：Yongji Kang and Huining Zong

指导教师单位：河南大学
Tutors from：Henan University

-DESIGN DESCRIPTION

此科技馆项目位于广东省广州市增城区，旨在打造一个现代化、生态化、智能化的科技交流与体验平台，集科技展览、创新互动、科普教育为一体，以推动广东科技文化发展，促进经济社会进步。

广东科技馆项目占地面积约30000平方米，建筑面积约25000平方米，采用现代科技元素和自然生态方式相结合的设计理念，注重通风遮阳，适应广州潮湿、炎热的地域气候特征。花形平面恰似广州市市花木棉花，花形平面配合流线型幕墙和均匀分布的拔风中庭为科技馆的使用人群创造了一个舒适、绿色的文化教育空间。

This science and technology museum project is located in Zengcheng District, Guangzhou City, Guangdong Province. It aims to create a modern, ecological and intelligent platform for scientific and technological exchange and experience, integrating scientific and technological exhibitions, innovation and interaction, and popularization of science and technology education, in order to promote the development of science and technology and culture in Guangdong, and to facilitate economic and social progress.

Covering an area of about 30 000 square meters and a building area of about 25 000 square meters, the Guangdong Science and Technology Museum project adopts a design concept that combines modern technological elements and natural ecological approaches, focusing on ventilation and shading and adapting to Guangzhou's humid and hot regional climate characteristics. The flower-shaped plan resembles the Guangzhou city flower Cotton Tree, which together with the streamlined curtain wall and the evenly distributed wind-pulling atrium creates a comfortable and green cultural and educational space for the users of the Science and Technology Museum.

1 Roof solar panel
2 Roof garden
3 Landscape pools
4 Atrium
5 wooden walkways
6 Parking Lot

Economic and technical indicators:
Site area:11447 m²
Base area:3977 m²
Total building area:12568 m²
Building density:34.74%
Plot ratio:1.09
Greening rate:32%
Height:24.00m

SITE PLAN 1:500

LOCATION ANALYSIS

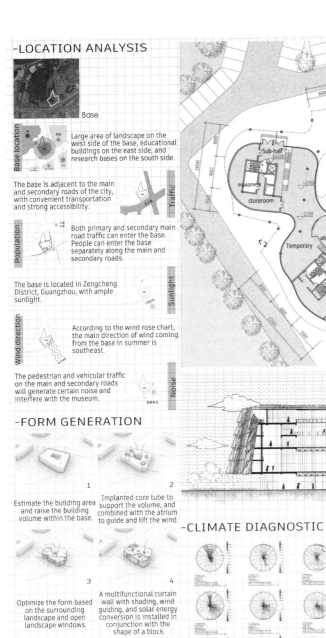

Base location: Large area of landscape on the west side of the base, educational buildings on the east side, and research bases on the south side.

Traffic: The base is adjacent to the main and secondary roads of the city, with convenient transportation and strong accessibility.

Population: Both primary and secondary main road traffic can enter the base. People can enter the base separately along the main and secondary roads.

Sunlight: The base is located in Zengcheng District, Guangzhou, with ample sunlight.

Wind direction: According to the wind rose chart, the main direction of wind coming from the base in summer is southeast.

Noise: The pedestrian and vehicular traffic on the main and secondary roads will generate certain noise and interfere with the museum.

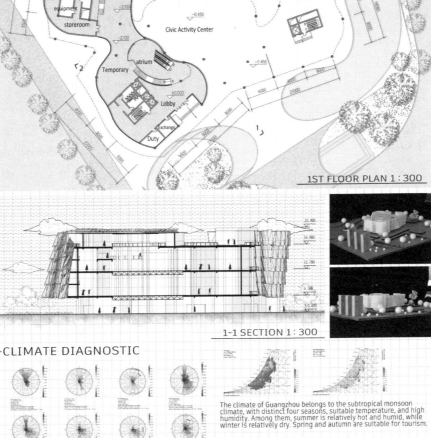

1ST FLOOR PLAN 1:300

1-1 SECTION 1:300

FORM GENERATION

1. Estimate the building area and raise the building volume within the base.
2. Implanted core tube to support the volume, combined with the atrium to guide and lift the wind.
3. Optimize the form based on the surrounding landscape and open landscape windows.
4. A multifunctional curtain wall with shading, wind guiding, and solar energy conversion is installed in conjunction with the shape of a block.

BUILT ON STILTS

Based on the location, a large area of overhead space on the ground floor has been selected for treatment. Firstly, it can promote ventilation and meet the wind demand in each room; The second is the consideration of the surrounding environment. A large area of gray space can form a civic activity square, even when the building is not open or people do not enter, it can bring vitality to the building.

CLIMATE DIAGNOSTIC

The climate of Guangzhou belongs to the subtropical monsoon climate, with distinct four seasons, suitable temperature, and high humidity. Among them, summer is relatively hot and humid, while winter is relatively dry. Spring and autumn are suitable for tourism.

In meteorology, the wind direction in Guangzhou is mainly influenced by the pressure fields in the South China Sea and the Asian continent, resulting in variable wind directions, mainly divided into southeast wind, northeast wind, southwest wind, etc. Among them, the southeast monsoon is prevalent in summer, and thunderstorms often occur in the evening; When the northern cold wave strikes in winter, it will bring cool northerly winds and may bring snow or cooling weather.

The precipitation in Guangzhou is mainly concentrated from May to September, with the rainy season from mid May to the end of June, with the highest precipitation in June. July to August is the peak of summer, with frequent thunderstorms and short-term heavy rainfall. In winter, it is cold in the morning and evening, warm during the day, with less rainfall and relatively dry.

SOUTH FACEDE   WEST FACEDE

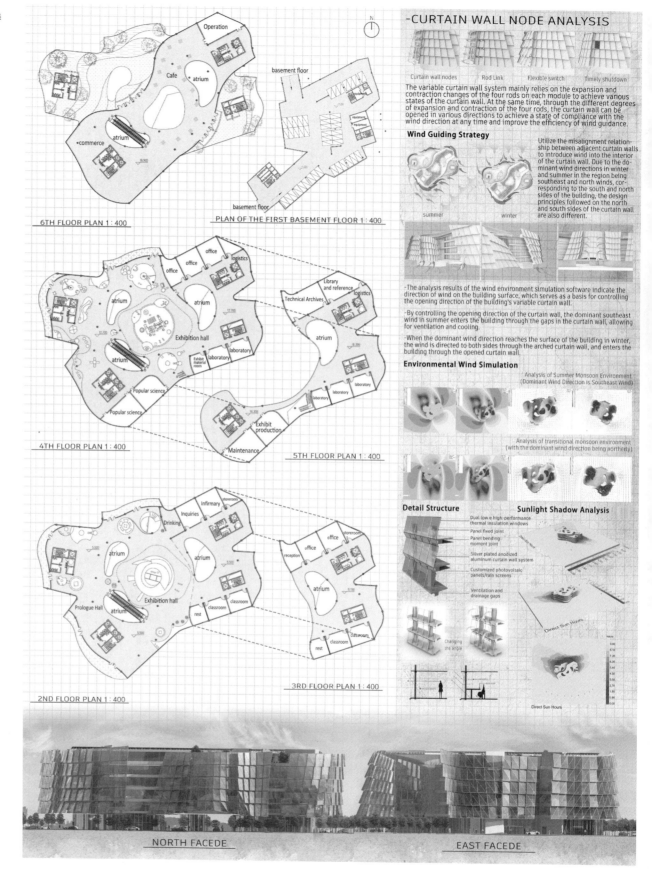

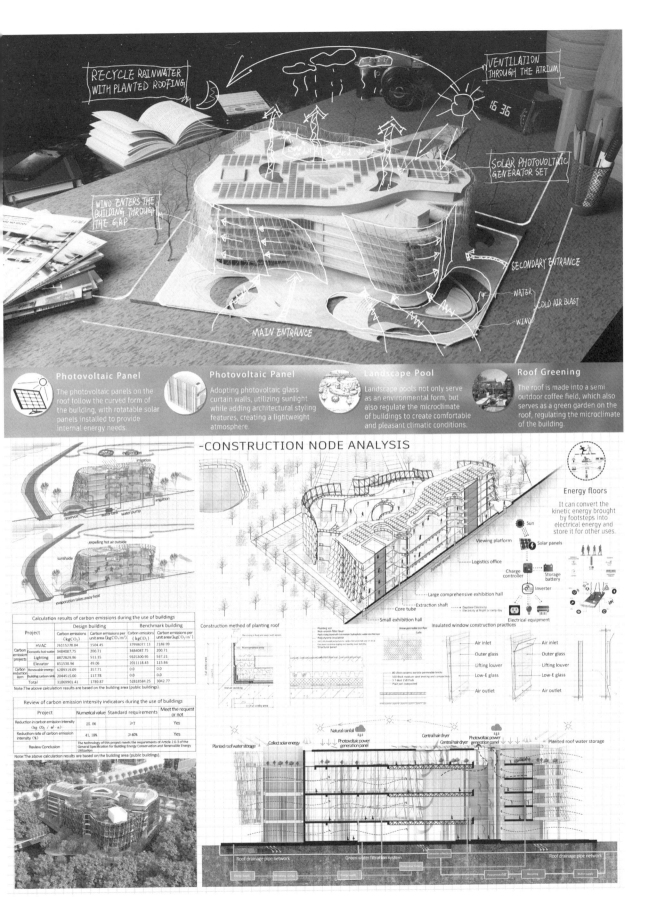

综合奖·优秀奖·零碳提升项目
Comprehensive Awards · Honorable Mention · Zero-Carbon Promotion Project

注册号：101320
Register Number：101320

项目名称：庭间序
Entry Title：Between Cavaedium

作者：杨林、卢一迪、刘静文、胡博雅、刘冰
Authors：Lin Yang, Yidi Lu, Jingwen Liu, Boya Hu and Bing Liu

作者单位：新疆大学、西安建筑科技大学、哈尔滨工业大学、同济大学
Authors from：Xinjiang University, Xi'an University of Architecture and Technology, Harbin Institute of Technology and Tongji University

指导教师：王万江、何泉、孟琪
Tutors：Wanjiang Wang, Quan He and Qi Meng

指导教师单位：新疆大学、西安建筑科技大学、哈尔滨工业大学
Tutors from：Xinjiang University, Xi'an University of Architecture and Technology and Harbin Institute of Technology

■ Location and Urban History

■ Scheme Design Specification

本着阳光·零碳的宗旨，我们对既有建筑进行了改造，我们希望在达到建筑零碳目标的同时，也能够改善空间的舒适度和空间趣味性，通过太阳能产能和植物固碳两个系统进行零碳改造。屋顶太阳能板以及立面的光伏玻璃作为主要的太阳能产能系统，中庭植入的"立体植物盒子"一方面采用光伏玻璃产能，另一方面进行植物固碳，处理了室内人员聚集场所产生的大量二氧化碳，同时还布设了水循环系统和除湿系统，分别进行雨水收集/中水回用和室内舒适度改造。

In line with the principle of sunlight and zero carbon, we renovated the existing building. We want to achieve the goal of zero carbon buildings, but also improve the space Between comfort and space of interest, through solar capacity and plant solid Carbon two systems carry out zero -carbonretrofits, rooftop solar panels as well as facades Photovoltaic glass as the main solar energy production system, the atrium is implanted. The "three-dimensional plant box" on the one hand uses photovoltaic glass production capacity, on the other hand Plant carbon sequestration was carried out to deal with the large amount produced in indoor people gathering places Carbon dioxide. At the same time, the water circulation system and dehumidification system are also arranged Rainwater collection/reuse of reclaimed water and indoor comfort renovation were carried out respectively.

■ Annual Wind Rose Chart

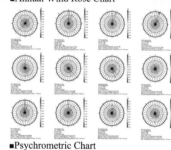

■ Psychrometric Chart

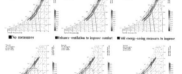

■ Atrium General Plan 1∶700

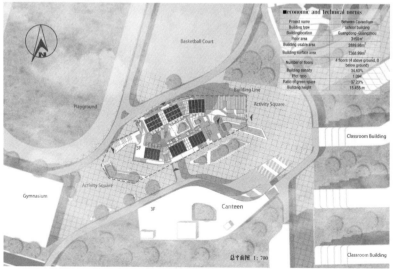

■ Energy Consumption Calculation Results

Project	Design building		Reference building	
	Total energy consumption (kW·h/a)	Energy consumption per unit area [kW·h/(m²·a)]	Total energy consumption (kW·h/a)	Energy consumption per unit area [kW·h/(m²·a)]
Heating energy consumption	1468.42	0.53	6355.76	2.29
Cooling energy consumption	369554.32	133.02	440443.32	158.54
Energy consumption of transmission and distribution system	206251.29	74.24	206251.29	74.24
Energy consumption of domestic hot water	57516.45	20.70	57516.45	20.70
Lighting system energy consumption	125220.85	46.14	128168.62	46.14
Elevator system energy consumption	8715.61	3.14	8715.61	3.14
Renewable energy generation	826565.40	86.57	-	-
Comprehensive value of building energy consumption excluding renewable energy generation	762736.94	182.83	847451.05	305.05
Comprehensive value of building energy consumption	-62828.46	138.99	847451.05	305.05

Note：1. The above calculation results are based on the building area.
2．When the comprehensive value of building energy consumption is negative, it indicates that the energy production of the building itself and surrounding renewable energy is greater than the annual terminal energy consumption, meeting the requirements of zero energy consumption building.

■ Explosion Analysis Diagram

太阳能光伏系统 / Solar photovoltaic system

The solar photovoltaic system can adjust the solar panel Angle according to the change of the sun Angle.
Solar panels convert solar energy into electricity and are the best choice for reducing carbon emissions.

植物固碳 / Sequestration box

We create functional boxes that people can use, while hanging plants around the boxes, adding interest to the building while neutralizing the building's carbon emissions.

爆炸图 / Explosive View

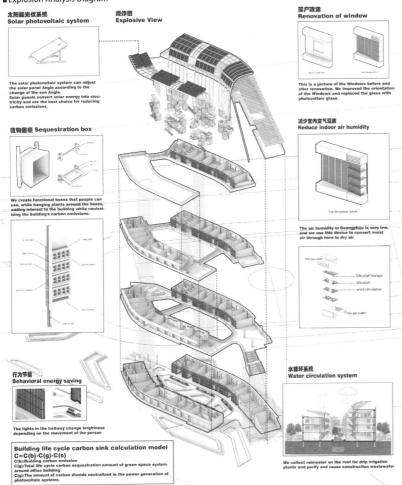

行为节能 / Behavioral energy saving

The lights in the hallway change brightness depending on the movement of the person.

Building life cycle carbon sink calculation model

$$C = C(b) - C(g) - C(s)$$

C(b): Building carbon emission
C(g): Total life cycle carbon sequestration amount of green space system around office building
C(s): The amount of carbon dioxide neutralized in the power generation of photovoltaic systems

窗户改造 / Renovation of window

This is a picture of the Windows before and after renovation. We improved the orientation of the Windows and replaced the glass with photovoltaic glass.

减少室内空气湿度 / Reduce indoor air humidity

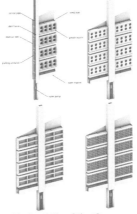

The air humidity in Guangzhou is very low, and we use this device to convert moist air through here to dry air.

水循环系统 / Water circulation system

We collect rainwater on the roof for drip irrigation plants and purify and reuse construction wastewater.

■ Fabricated Plant Petri Dishes

Water pumping can promote the operation of water circulation system during the building operation, so that water can circulate in the building. This assembled three-dimensional plant module can cultivate plants under the action of water flow. The plant box can enrich the space function, and at the same time, plant carbonfixation can promote aero-carbon construction.

■ The Possibility of Plant Boxes

The following shows the spatial plasticity of the functional boxes we designed,showing several different spatial possibilities, enriching the space of the community complex building and increasing the space interest.

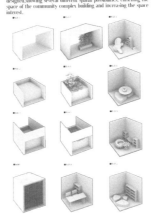

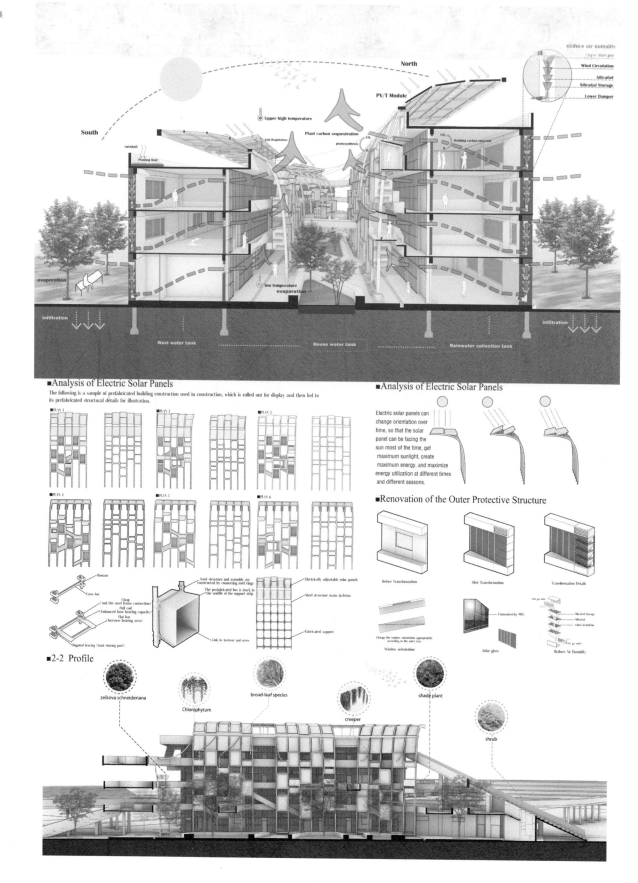

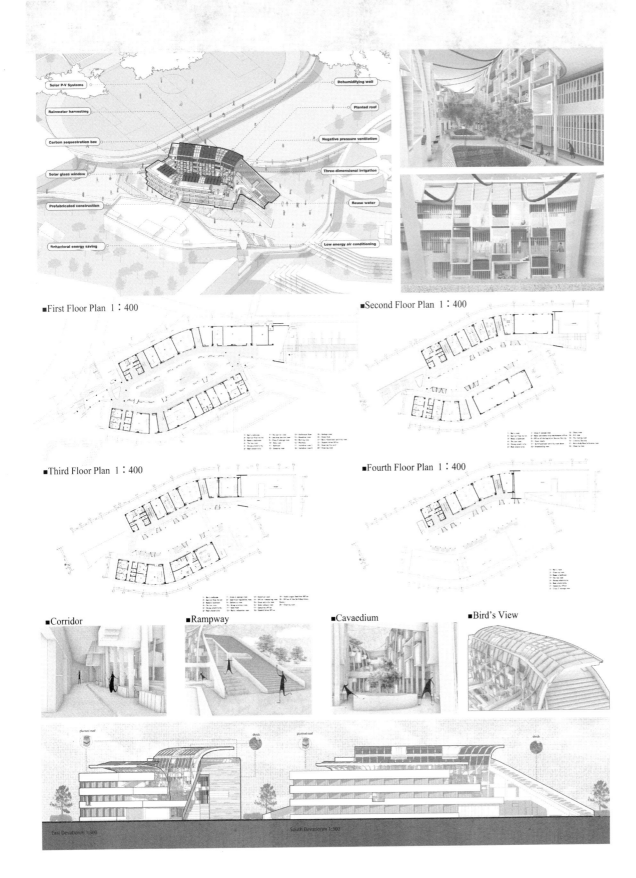

综合奖·优秀奖·零碳提升项目
Comprehensive Awards · Honorable Mention · Zero-Carbon Promotion Project

注册号：101522
Register Number：101522

项目名称：从游·木筑
Entry Title：Legolas Rotating

作者：吴颖怡、方瑶璇、杜晟、卢潇月
Authors：Yingyi Wu, Yaoxuan Fang, Sheng Du and Xiaoyue Lu

作者单位：东北大学
Authors from：Northeast University

指导教师：刘哲铭
Tutor：Zheming Liu

指导教师单位：东北大学
Tutor from：Northeast University

□ 设计说明 / Design Instructions

基地位于位于广州市增城区广州科教城一期核心地带西侧的广州市公用事业技师学院内，西邻运动场，南邻南食堂，东邻校园主干道。根据基地周边现状以及未来建筑的功能要求，对其进行设计，旨在突出建筑的前瞻性和实用性，顺应当下绿建趋势，内置多种绿色技术、雨水收集技术、太阳能光热光电技术、屋顶和立面绿化、呼吸幕墙等，打造低消耗、低排放的节能建筑。

The base is located in Guangzhou Public Utilities Advanced Technical School, located in the west of the core area of Guangzhou Science and Education City Phase I, Zengcheng District, Guangzhou, adjacent to the sports field in the west, the South Canteen in the south, and the main campus road in the east. According to the current situation around the base and the functional requirements of future buildings, it is designed with the concept of "power source a vortex", aiming to highlight the forward-looking and practical nature of the building, conform to the trend of green construction, and build a variety of green technologies: rainwater harvesting technology, solar thermal photovoltaic technology, roof and façade greening, breathing curtain wall, etc.

□ 改造策略 / Retrofit Strategy

在对建筑现状最大程度的合理保留下，思考如何处理建筑立面与屋顶，实现降低建筑用能需求和碳排放，集成太阳能等可再生能源系统，实现建筑全年用能与产能平衡。
南立面：根据日照朝向，固定式遮阳构件与太阳能板竖向结合转向布置。
东立面：上面露台遮阳与太阳能光伏技术结合布置，激活室外场地活力。
西立面：攀爬架式绿化遮阳，降低太阳直射影响，减少西晒不利影响。
北立面：采用绿化悬挂的设计策略，延续生态屋面，展现校园主要风貌。
第五立面：木构遮阳光伏一体化设计，配合屋顶绿化，传递绿色理念。

□ 改造策略图解 / Illustrations of Retrofit Strategy

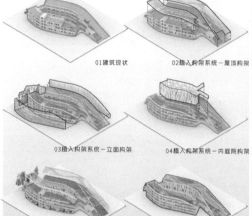

□ 场地环境分析 / Site Location Analysis

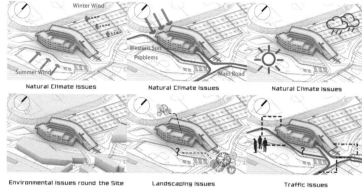

□ 总平面图 / General Layout

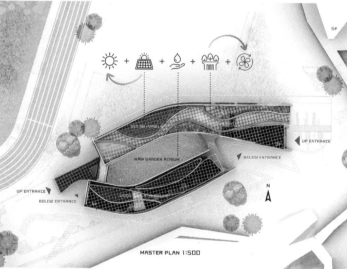

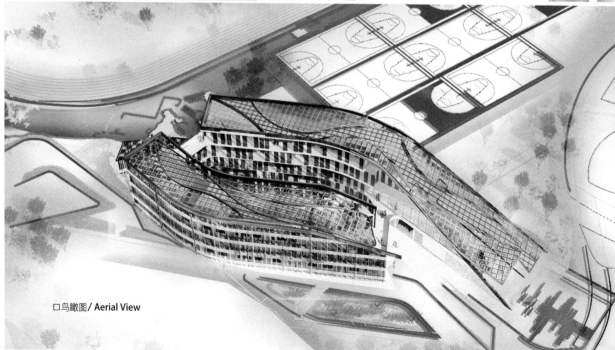

□ 鸟瞰图 / Aerial View

气候分析/Climate Analysis

汇总/summary

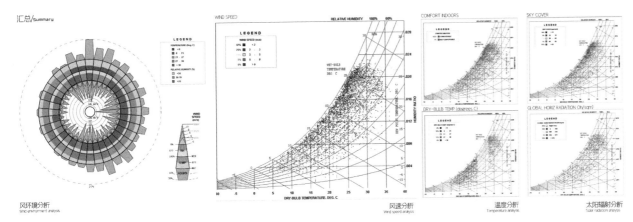

风环境分析 / Wind environment analysis　　风速分析 / Wind speed analysis　　温度分析 / Temperature analysis　　太阳辐射分析 / Solar radiation analysis

各层平面图/Floor Plans

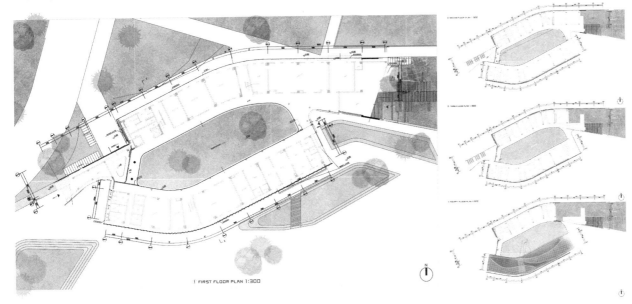

1 FIRST FLOOR PLAN 1:300

中庭改造策略分析/Analysis of Atrium Renovation Strategy

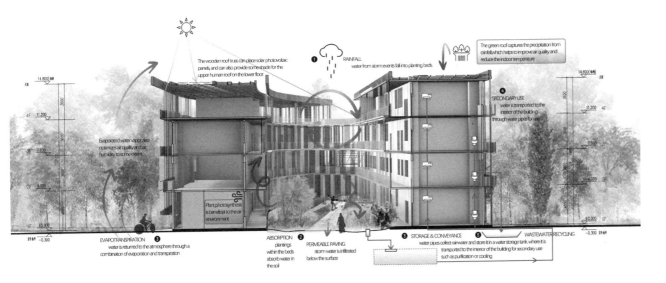

□ 立面绿化系统/Livingwalls

□ 植物配置分析/Plant Configuration Analysis

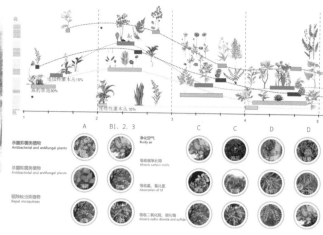

□ 南立面改造分析/Analysis of the Renovation of the South Facade

□ 植物配置分析/Plant Configuration Analysis

□ 绿化构造节点/Greening Construction Node

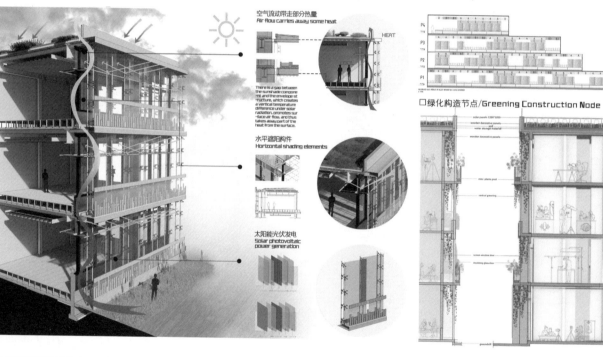

□ 中庭改造策略分析/Analysis of Atrium Renovation Strategy

1-1 剖面图 1:300　　1-1 CUTAWAY 1:300

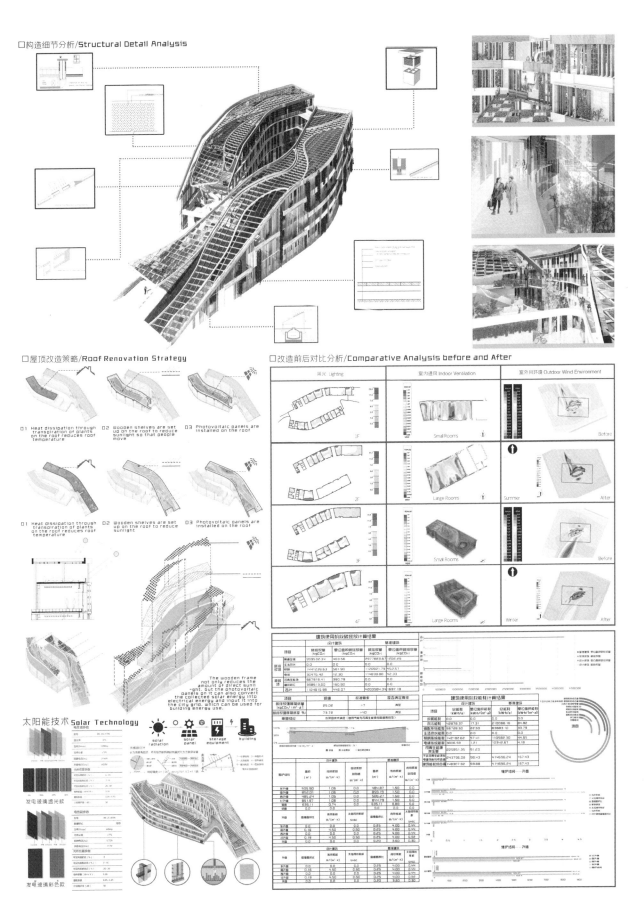

综合奖·优秀奖·零碳提升项目
Comprehensive Awards · Honorable Mention · Zero-Carbon Promotion Project

注册号：101774
Register Number：101774

项目名称：风廊·叶影
Entry Title：Wind Corridor, Leaf Shadow

作者：蒋辰瑜、彭榆棋、蒙约
Authors：Chenyu Jiang, Yuqi Peng and Yue Meng

作者单位：重庆大学、东南大学
Authors from：Chongqing University and Southeast University

指导教师：周铁军、谢崇实、张海滨
Tutors：Tiejun Zhou, Chongshi Xie and Haibin Zhang

指导教师单位：重庆大学、重庆设计集团有限公司
Tutors from：Chongqing University and Chongqing Design Group Co., Ltd.

Site Analysis

Design Description

本方案以"风廊叶影"为主题，同时包括了方案的两大被动技术核心，即拔风走廊和可变的遮阳百叶。将走廊外扩形成两个双层表皮的拔风烟囱，与各房间联通形成整体的被动式通风系统，底层加强得堂风。同时在立面进阳、屋顶绿化等方面进行了因地制宜的被动节能设计。

为了实现全年运行零碳排放，方案通过软件模拟，对太阳能光伏板的铺设角度进行了精确的计算。经过软件计算，可以达到全年101.2%的碳中和率。

This plan focuses on the theme of "Wind Corridor and Leaf Shadow", including the two main passive technical cores of the plan, namely the wind corridor and variable sunshade blinds. Expanding the corridor to form two double skin exhaust chimneys, connecting them with each room to form a passive ventilation system as a whole, and strengthening ventilation at the bottom. At the same time, passive energy-saving design has been carried out in areas such as facade shading and roof greening that are tailored to local conditions.

In order to achieve zero carbon emissions throughout the year, the plan accurately calculated the laying angle of solar photovoltaic panels through software simulation. After software calculation, it can achieve a carbon neutrality rate of 101.2% for the entire year.

Monthly Wind Rose Chart

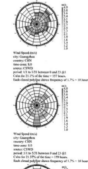 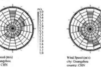   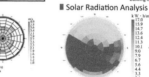

Solar Radiation Analysis

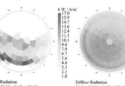

Summer&Winter Psychrometric Chart

Annual Temperature Statistics

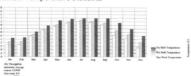 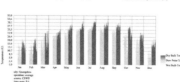 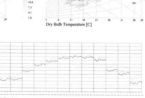

■ First Floor Plan 1∶250

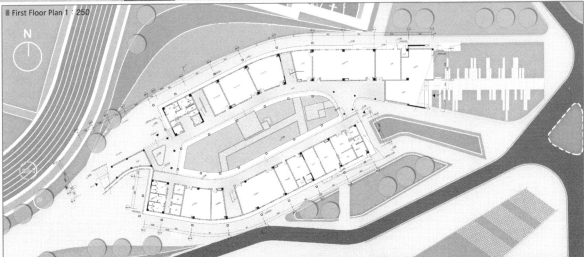

■ Zero Energy Building Strategy Map

■ Cultural Symbols Analysis

Arcade
In order to create cool and comfortable living conditions, they used the method of adding corridors in front of the living rooms to avoid the heat.

Symbol application
While designing the green building concept, we also considered the echo of the local cultural symbol - the arcade. The construction plan with the additional she-if as the core not only co-ordinates the overall plan, but also produces the same limited space pattern as the arcade in the first floor space.

Gray space
Gray space is formed by the limitation of columns and walls.

■ Material Analysis

Antiseptic walnut   Light wood panels   Low-e glass   Photovoltaic panels

■ Behavioral Analysis

■ South Elevation 1∶250   ■ East Elevation 1∶250

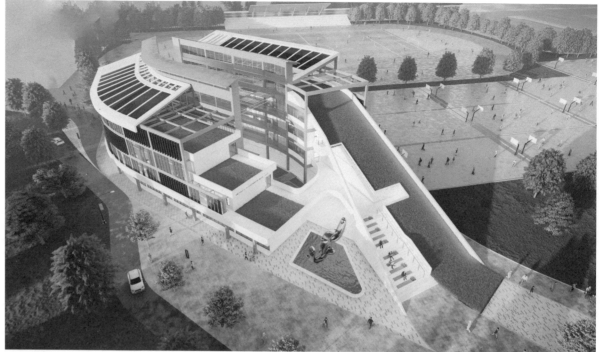

■ Plan View

■ Exploded View

■ Wall Detail

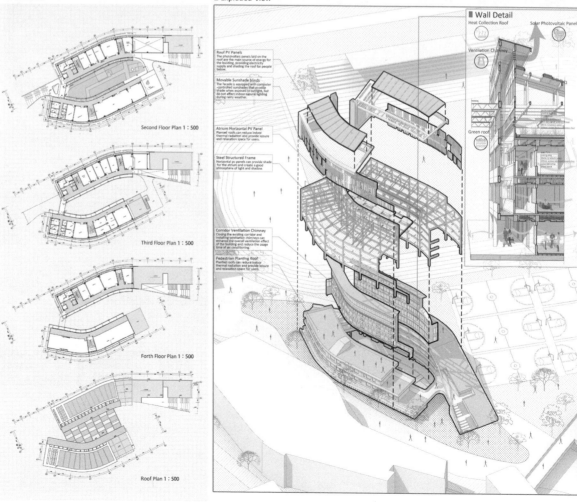

Second Floor Plan 1 : 500

Third Floor Plan 1 : 500

Forth Floor Plan 1 : 500

Roof Plan 1 : 500

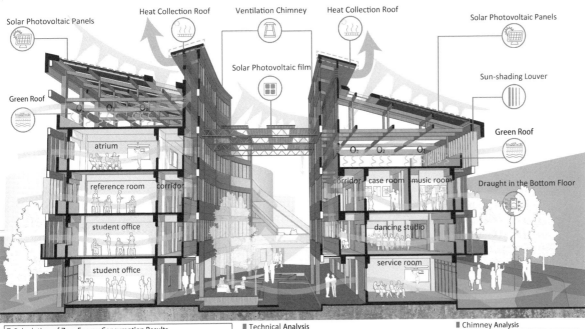

Calculation of Zero Energy Consumption Results

Sheet 1: Energy consumpion proportion
Sheet 2: Monthly PV power generation

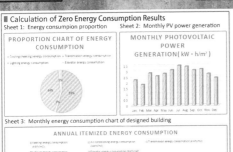

Sheet 3: Monthly energy consumption chart of designed building

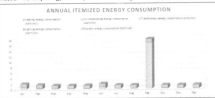

Sheet 4: Energy consumption calculation results

PROJECT	Designed building		Benchmark building	
	Total energy consumption (kW·h/a)	Energy consumption per unit area [kW·h/(m²·a)]	Total energy consumption (kW·h/a)	Energy consumption per unit area [kW·h/(m²·a)]
Heating energy consumption	2329.93	0.95	3213.4	1.31
Cooling energy consumption	42098.28	17.18	43733.28	17.85
Transmisson energy consumption	61.16	0.02	15358.04	6.27
Lighting energy consumption	35301.02	14.41	76793.37	31.34
Elevator energy consumption	1232.33	0.5	4138.25	1.69
Renewable energy generation	84205.9	34.36	-	-
Comprehensive value of building energy consumption excluding renewable energy generation	81022.72	33.06	143236.35	58.45
Comprehensive value of building energy consumption	-1706.5	-0.7	143236.35	58.45
Conclusion	The comprehensive value of building energy consumption is negative, and the energy production of the building itself and surrounding renewable energy is greater than the annual terminal energy consumption of the building, meeting the requirements for zero energy consumption buildings.			

Sheet 5: Review of building energy consumption indicators

Review of energy efficiency indicators for buildings with near zero energy consumption			
Project	Number	Standard requirements	Satisfied or Not
Comprehensive energy efficiency (%)	101.19	≥60	Satisfied
Utilization of renewable energy (%)	103.93	≥10	Satisfied
Architectural noumenon Energy efficiency rate of building (%)	43.43	≥20	Satisfied
Performance index Air changes N₅₀	0.5	≤-	Satisfied
Conclusion	The technology of this project meets the requirements of near zero energy consumption buildings in the "Technical Standards for Near Zero Energy Consumption Buildings".		

Technical Analysis

The Chimney Effect | Bottom floor ventilation | Atrium ventilation effect

Sunshade on sunny days | Transparent on cloudy days | Turn on at night to promote ventilation

Photovoltaic panels avoid direct sunlight on the roof | PV panels provide electricity for lighting and air conditioning | The wind from the chimney carries away the heat from the PV panel

Ventilation Corridor Air Flow Organization | Atrium airflow organization | Top and bottom airflow organization

Construction Steps

Step 1: Erecting steel frames outside the building | Step 2: Close the corridor and set up ventilation chimney outside

 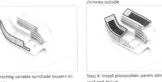 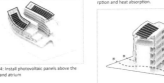

Step 3: Erecting variable sunshade louvers on steel frames | Step 4: Install photovoltaic panels above the roof and atrium

1-1 Section 1 : 250

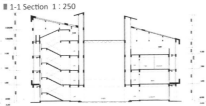

Chimney Analysis

The ventilation chimney is composed of a double-layer skin, and the size of the chimney is determined to be reasonable through calculation

The ventilation chimney is composed of glass windows both inside and outside, without obstructing the external view. The inner window sash can be opened and closed independently.

The ventilation chimney is constructed of darker colored materials, which promote the ventilation effect after light absorption and heat absorption.

The bottom layer of the ventilation chimney is open, promoting the ventilation effect while creating facade elements.

Reserve a gap between the ventilation chimney and the PV system to facilitate the organization of ventilation airflow.

综合奖·优秀奖·零碳提升项目
Comprehensive Awards · Honorable Mention · Zero-Carbon Promotion Project

注册号：101843
Register Number：101843

项目名称：还碳·环碳
Entry Title：Carbon Returning and Carbon Cycling

作者：林裳、沈鑫、赵雅雯、梁心
Authors：Chang Lin, Xin Shen, Yawen Zhao and Xin Liang

作者单位：内蒙古工业大学
Authors from：Inner Mongolia University of Technology

指导教师：伊若勒泰、许国强
Tutors：Yiruoletai and Guoqiang Xu

指导教师单位：内蒙古工业大学
Tutors from：Inner Mongolia University of Technology

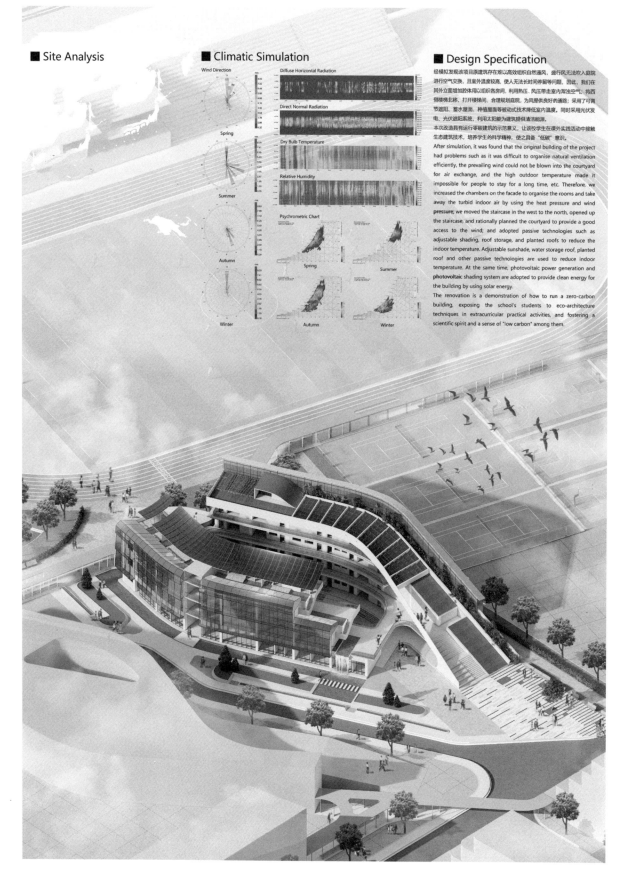

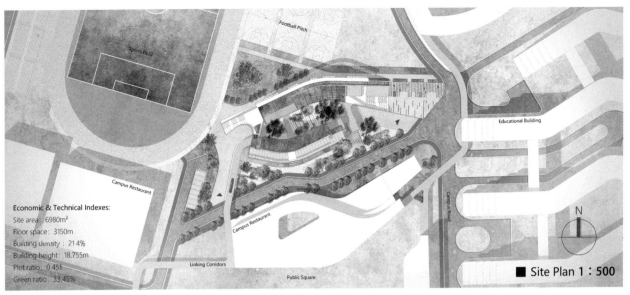

Economic & Technical Indexes:
Site area: 6980m²
Floor space: 3150m
Building density: 21.4%
Building height: 18.755m
Plot ratio: 0.451
Green ratio: 33.45%

■ Site Plan 1 : 500

■ Site Planning Analysis

① **Greening Analysis**
The surrounding area of the building and the campus are covered with greenery, including lawns, shrubs, and trees.

② **Perimeter Analysis**
The north and northwest sides of the building are sports venues, the south side is the cafeteria, the northeast is the library, and the southeast is the teaching building.

③ **Road analysis**
The road on the south side of the building leads directly to the main entrances of the east and west sides of the campus, and the building is located in the center of the road.

④ **Wind Environment Analysis**
The region is plagued by southeast winds throughout the year.

⑤ **Flow Analysis**
The surrounding area of the building is mainly for campus use, with a high pedestrian flow.

⑥ **Sunlight Analysis**
The region has a high solar altitude angle and a high amount of solar radiation.

⑦ **Building Height Analysis**
The dormitory, teaching building, and library have a height of 6 floors, the cafeteria has a height of 3 floors, and the building itself has a height of 4 floors.

⑧ **Noise Analysis**
The building is located in the center of the campus, and the noise impact of urban roads is not significant.

■ Program Generation

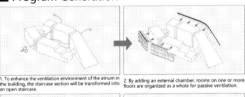

1. To enhance the ventilation environment of the atrium in the building, the staircase section will be transformed into an open staircase.
2. By adding an external chamber, rooms on one or more floors are organized as a whole for passive ventilation.

3. Using water storage and planting roofs on the roof for cooling.
4. Install solar photovoltaic panels on the roof to provide shade and insulation while producing capacity.

5. The stairs on the west side of the building block the summer prevailing winds.
6. Move the stairs northward to expose the atrium space, and use trees to guide the wind into the atrium.

7. Adapt to the architectural and site forms, introduce flooring and water bodies.
8. Finally, following the architectural trend, introducing flower beds to enhance the richness and completeness of the site.

■ Analysis of Technical Routes

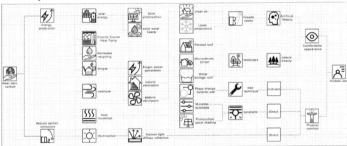

■ Technology Mapping

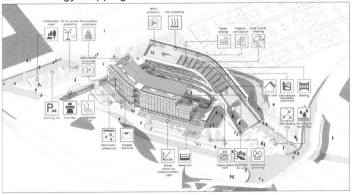

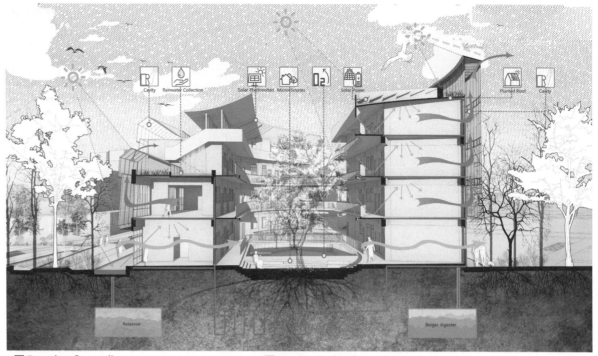
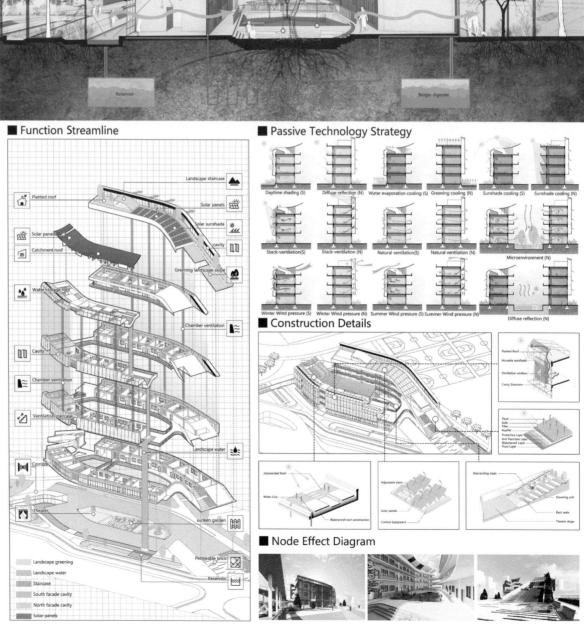

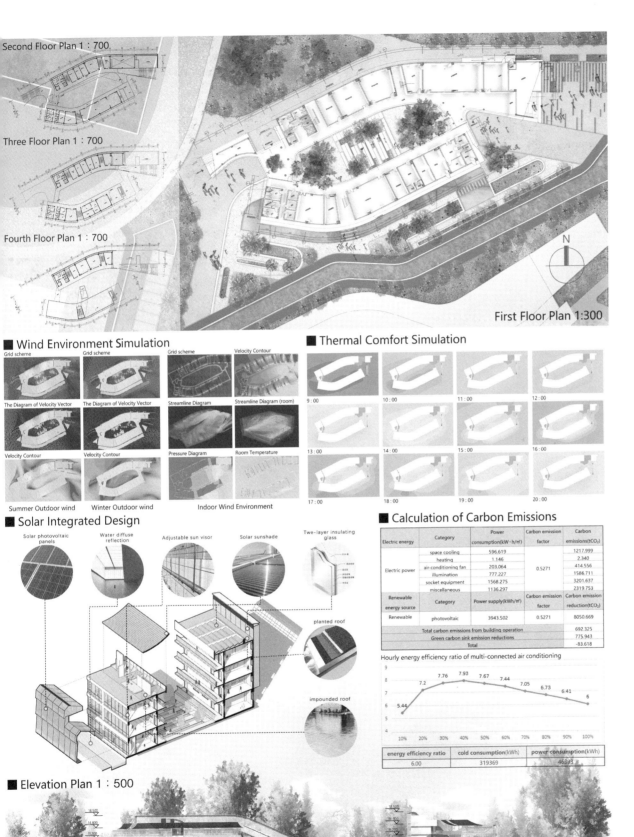

综合奖·优秀奖·零碳提升项目
Comprehensive Awards · Honorable Mention · Zero-Carbon Promotion Project

注册号：101899
Register Number：101899

项目名称：翕风纳绿
Entry Title：Delightful Wind Brings Greenery

作者：罗璘曦、韦海璐、郭家玥、角元昊、王岚、陈文颀、王颖旭
Authors：Linxi Luo, Hailu Wei, Jiayue Guo, Yuanhao Jue, Lan Wang, Wenqi Chen and Yingxu Wang

作者单位：东南大学
Authors from：Southeast University

指导教师：刘一歌、王伟
Tutors：Yige Liu and Wei Wang

指导教师单位：东南大学
Tutors from：Southeast University

Design Specification

考虑到广州充足的光热资源，通过气候适应性设计改善建筑的遮阳并加强通风是非常重要的。设计受到传统岭南建筑冷巷空间系统的启发，关注外立面、屋顶及中庭。通过立面与屋顶双层表皮设计，引入垂直的冷巷，加强拔风效应，并通过表皮的开合有效调节遮阳、通风，最终减少25.1%能耗。此外，增加光伏实现建筑产能210.47MW·h，结合庭院与屋顶绿化的碳汇作用，实现建筑全年的用能与产能平衡，实现建筑运行零碳排放，满足绿建标准的三星要求。

Considering Guangzhou's abundant light and heat resources, it is important to improve the building's shading and enhance ventilation through climate-adaptive design. The design is inspired by the traditional Lingnan architectural cold alley space system, focusing on the façade, roof and atrium. Through the double skin design of the façade and roof, a vertical cold alley is introduced to enhance the wind pulling effect, and the opening and closing of the skin effectively regulates shading and ventilation, ultimately reducing energy consumption by 25.1%. In addition, the addition of photovoltaics achieves a building capacity of 210.47MWh, which, combined with the carbon sink effect of the courtyard and roof greening, achieves a balance between the building's energy consumption and capacity throughout the year, and realizes zero carbon emissions from the building's operation, meeting the three-star requirements of the green building standard.

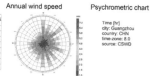

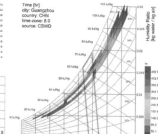

Climate Analysis

Guangzhou has abundant light and heat resources and abundant rainfall. The rainy season is from April to September, and the precipitation accounts for more than 80 per cent of the year. In addition, it has a prominent monsoon climate with warm and humid weather.

Birds View

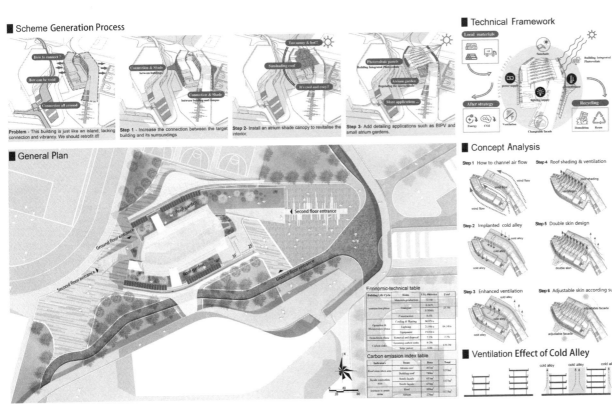

■ Typical Floor Plan 1:300

■ Building Profile (1:300)

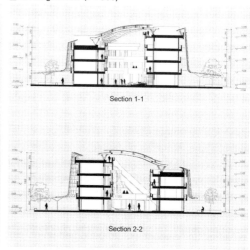

Section 1-1

Section 2-2

■ Double Roof Skin Analysis

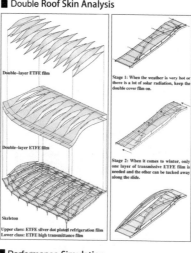

Double-layer ETFE film

Double-layer ETFE film

Skeleton

Upper class: ETFE silver dot plated refrigeration film
Lower class: ETFE high transmittance film

Stage 1: When the weather is very hot or there is a lot of solar radiation, keep the double cover film on.

Stage 2: When it comes to winter, only one layer of transmissive ETFE film is needed and the other can be tucked away along the slide.

■ Variable Facade Skin Analysis

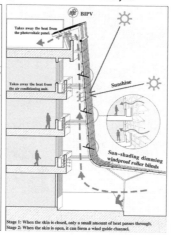

Stage 1: When the skin is closed, only a small amount of heat passes through.
Stage 2: When the skin is open, it can form a wind guide channel.

■ Ventilation Simulation Analysis

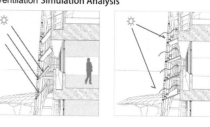

■ Performance Simulation

Wind velocity diagram

 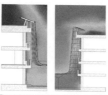

Air rises under the influence of hot pressure and finally exits through gaps in the roof, improving atrium ventilation.

The wind goes straight up along the gap between the wall and the visor, while carrying away the heat.

Emperature analysis diagram

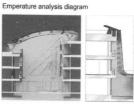

After analysis, the atrium and facade of this design have the effect of pulling wind

Green Building Evaluation Analysis

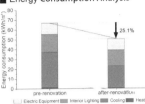

After renovation, overall green building score changed from **57.5** (2-star) to **80.32** (3-star).

Energy Consumption Analysis

Using Honeybee for energy simulation, the total energy consumption was reduced from 66.799kW·h/m² to 49.995kW·h/m² through renovation.

Specific Additions to the Score

Evaluation indicators	Section numbe	pre-score	after-score	Specific Improvement
Land saving & Outdoor environment	4.2.7	0	4	1. External sunshade, with a shade area of 1,127 m² and a shade ratio of >30%; 2. more than 70% of the road pavement, building roof solar radiation reflection coefficient is not less than 0.4;
	4.2.9	4	6	All buildings are connected by a storm corridor.
	4.2.15	3	6	The percentage of green roofs is 78 per cent.
Energy saving & Energy utilisation	5.2.2	0	6	50 per cent of openable area for external windows.
	5.2.3	0	6	The annual calculated load reduction for heating and air conditioning for the project is 24.65 per cent.
	5.2.4	0	6	The area covered by green roof measures has reached 78 per cent, and more than 80 per cent of the windows are effectively shaded.
	5.2.13	0	9	Electric fans are used in the main functional areas; external sun-shading measures are adopted for the east and west exterior windows and curtain walls of the buildings.
	5.2.16	0	10	The form of renewable energy used in this project is solar energy; the proportion of electricity supplied by renewable energy is greater than 80 per cent.
Material saving & Material resource utilisation	7.2.4	0	5	Reusable partitions (walls) are used in more than 80 per cent of the interior spaces in buildings with changeable functions.
	7.2.11	0	5	Highly durable building structure materials have been rationally used.
	7.2.14	0	8	Rational use of durable and easy-to-maintain decorative building materials.
Indoor environmental quality	8.2.5	0	3	The main functional rooms of the building have good outdoor views.
	8.2.8	0	12	The transparent parts of the project's external windows and curtain walls use adjustable external sunshading panels, with an area ratio of more than 80 per cent throughout the year.
bonus points	11.2.1	0	2	HVAC calculated load reductions of up to 15 per cent throughout the year.
	11.2.12	0	4	Carbon emission calculations and analyses have been carried out, and measures have been taken to reduce the carbon emission intensity per unit of floor area.

Building Life Cycle Analysis

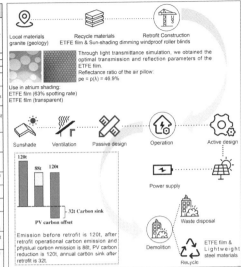

Local materials granite (geology) / Recycle materials ETFE film & Sun-shading dimming windproof roller blinds / Retrofit Construction

Through light transmittance simulation, we obtained the optimal transmission and reflection parameters of the ETFE film.
Reflectance ratio of the air pillow:
pe = p(λ) = 46.9%

Use in atrium shading:
ETFE film (63% spotting rate)
ETFE film (transparent)

Sunshade / Ventilation / Passive design / Operation / Active design
Power supply / Waste disposal / Demolition / Recycle

Emission before retrofit is 120t, after retrofit operational carbon emission and physical carbon emission is 88t, PV carbon reduction is 120t, annual carbon sink after retrofit is 32t.

Solar PV Panel Setup

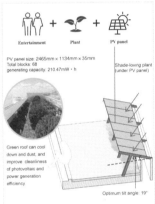

Entertainment + Plant + PV panel

PV panel size: 2465mm × 1134mm × 35mm
Total blocks: 68
generating capacity: 210.47mW·h

Shade-loving plant (under PV panel)

Green roof can cool down and dust, and improve cleanliness of photovoltaic and power generation efficiency

Optimum tilt angle: 19°

The addition of photovoltaic panels to the retrofit project not only fulfils the zero-carbon requirement for the operation and maintenance phase of the building, but also makes the building "carbon-negative".

Building North Elevation & Building South Elevation (1:300)

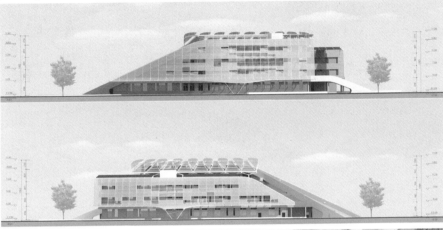

综合奖·优秀奖·零碳提升项目
Comprehensive Awards · Honorable Mention · Zero-Carbon Promotion Project

注册号：101997
Register Number：101997

项目名称：风谷之竹
Entry Title：Bamboos in the Valley

作者：徐浩晨、邵子沐、柴西妮、王飞、吕依涵
Authors：Haochen Xu, Zimu Shao, Xini Chai, Fei Wang and Yihan Lyu

作者单位：东南大学
Authors from：Southeast University

指导教师：伊若勒泰
Tutors：Yiruoletai

指导教师单位：内蒙古工业大学
Tutors from：Inner Mongolia University of Technology

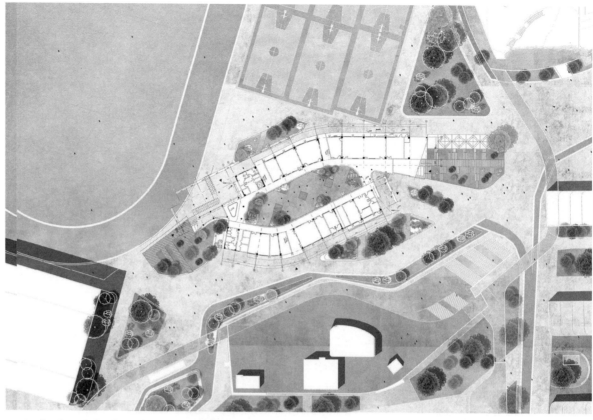

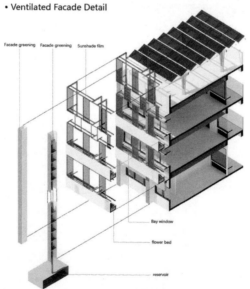
• Ventilated Facade Detail

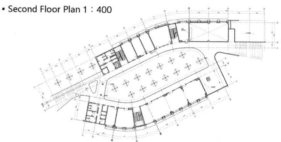
• Second Floor Plan 1：400

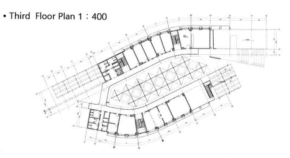
• Third Floor Plan 1：400

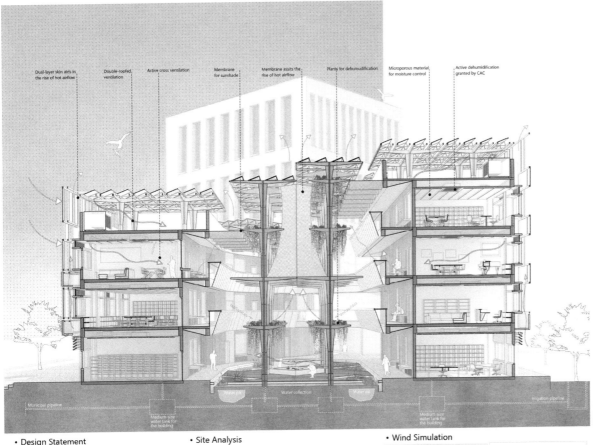

• Design Statement

Due to its humid subtropical climate and hilly terrain, the South China region has long been one of the major bamboo-producing areas. In recent years, with the gradual maturity of technology, bamboo materials have been increasingly integrated into construction components. The use of bamboo not only brings a soothing warm tone to the dull concrete structures but also offers a sustainable choice due to its inherent carbon sequestration capabilities and profound cultural significance.

The core focus of this project revolves around creating a suitable campus environment within the existing architecture. Leveraging the adaptability of the central courtyard space, the project introduces standardized bamboo structures as "green microelements" into the existing buildings. These structures serve multiple functions such as rainwater collection, ventilation, dehumidification, shading, and energy generation. Accompanied by facade renovations, the aim is to transform the building into a physically comfortable, ecologically suitable, calming, and vibrant space for various activities.

• Design Concept

• Technological Strategy

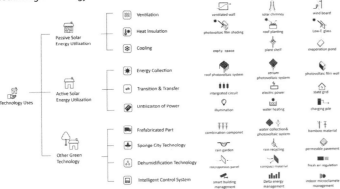

• Site Analysis

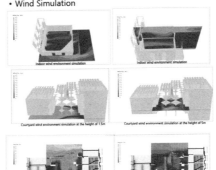

Guangzhou is characterized by relatively warm temperature throughout the year due to its lower latitude. The annual average temperature ranges from 20 to 22 degrees Celsius. During the winter months, Guangzhou is primarily influenced by northward winds, with a prevalence of northeast winds. In contrast, during the summer, the prevailing winds are from the south, with a higher occurrence of southeast winds. Humidity, sunshade and cooling is a major concern of this site

• Wind Simulation

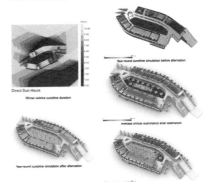

• Solar Simulation

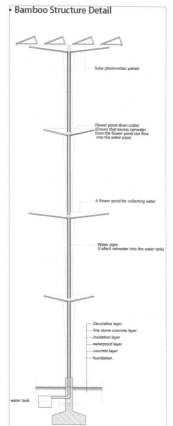

• Bamboo Structure Detail

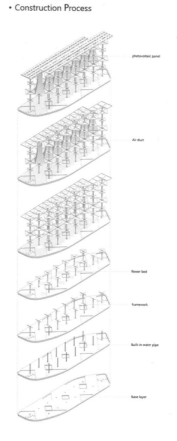

• Construction Process

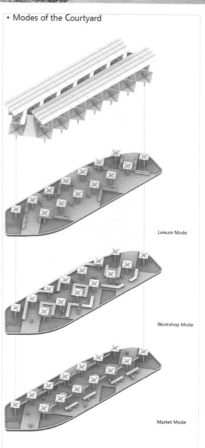

• Modes of the Courtyard

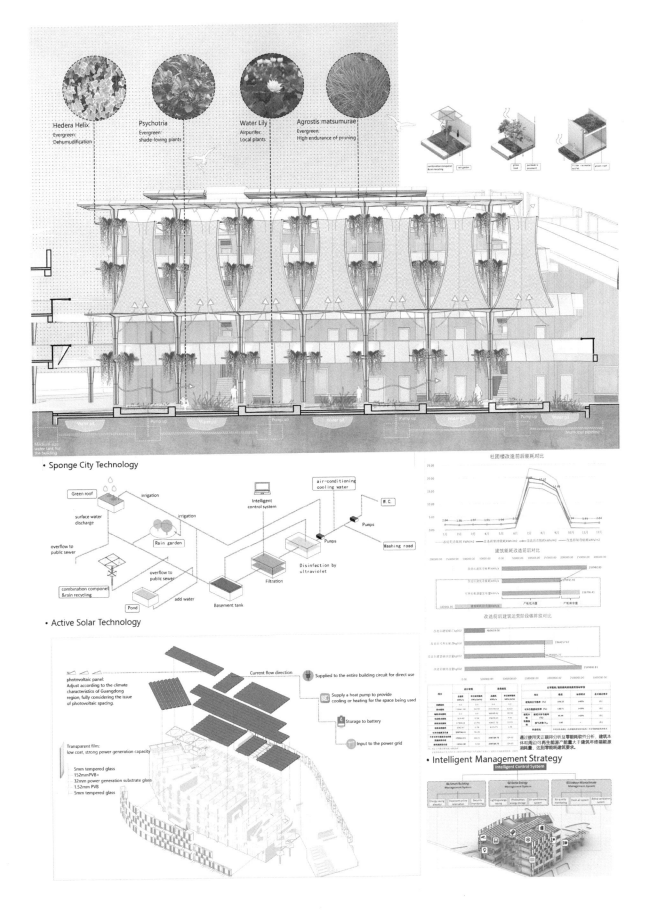

综合奖·入围奖·零碳设计项目
Comprehensive Awards · Nomination · Zero-Carbon Design Project

注册号：101304
Register Number：101304

项目名称：天工开物
Entry Title：Buildings Torn Apart by Plants

作者：田晓可、张昕岩、肖筱依、赖哲航
Authors：Xiaoke Tian, Xinyan Zhang, Xiaoyi Xiao and Zhehang Lai

作者单位：厦门大学
Tutors from：Xiamen University

指导教师：贾令堃、石峰
Tutors：Lingkun Jia and Feng Shi

指导教师单位：厦门大学
Tutors from：Xiamen University

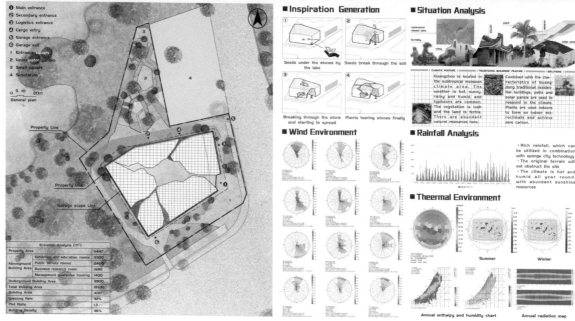

■ Block Generation

■ Spatial Sequence

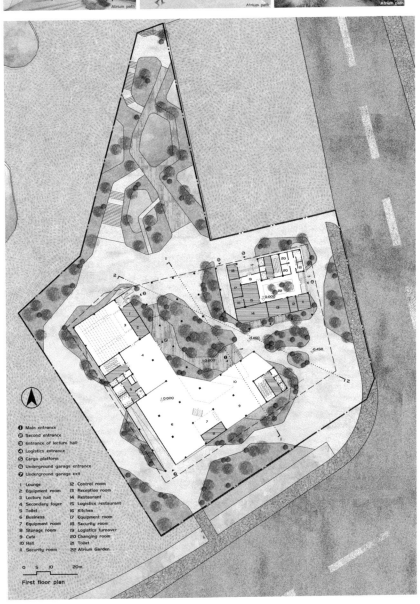

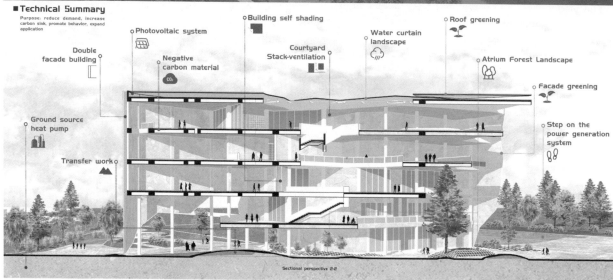

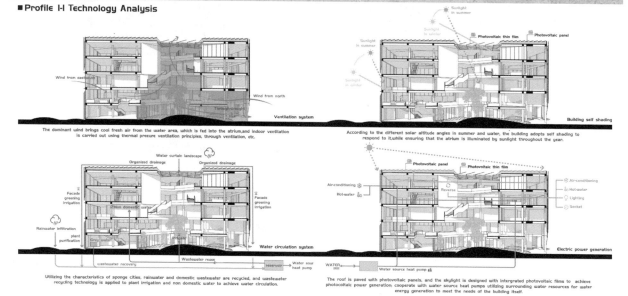

■ Explosion Diagram

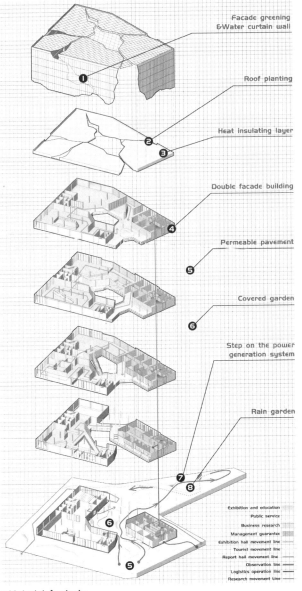

1. Facade greening & Water curtain wall
2. Roof planting
3. Heat insulating layer
4. Double facade building
5. Permeable pavement
6. Covered garden
7. Step on the power generation system
8. Rain garden

- Exhibition and education
- Public service
- Business research
- Management guarantee
- Exhibition hall movement line
- Tourist movement line
- Report hall movement line
- Observation line
- Logistics operation line
- Research movement line

■ Energy Consumption and Carbon Emissions Calculation

Building carbon emission (Use phase)		
Projects		Carbon emission
Carbon emission terms	HVAC	7129121.26
	Domestic hot water	380000.00
	Illumination	1749490.75
	Elevator	59469.45
	Total	9318081.46
Carbon reduction terms	Renewable energy	8722667.72
	Building carbon sink	2999880.00
	Total	11722547.72
Total		0.00

Energy consumption calculation:
In the building use stage, the annual energy consumption is 881768.61 kW·h, and the annual energy output is 104,7017.9 kW·h, which can achieve the annual energy consumption balance. Among them, capacity items include photovoltaic power generation and stampede power generation systems.

Carbon emission is calculated according to AutoCAD. The calculation results are shown in the table on the left.

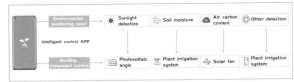

■ Intelligent Control System

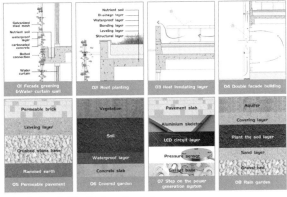

■ Structure Analysis (Corresponding Explosion Diagram)

01 Facade greening & Water curtain wall
02 Roof planting
03 Heat insulating layer
04 Double facade building
05 Permeable pavement
06 Covered garden
07 Step on the power generation system
08 Rain garden

■ Radiation & Ventilation Simulation

SUMMER | WINTER | 2M | 6M | 12M | 18M

■ Plant Analysis

Facade greening:
Beautify the environment, return to nature, dust and noise reduction, environmental protection, make indoor winter warm and summer cool, create a healthy and natural living environment, and improve the urban heat island effect.

Atrium greening:
By blocking solar radiation from plants, it creates a comfortable thermal environment for the semi outdoor space of the building, regulates the microclimate of the atrium, and serves as an important landscape inside the building.

■ Material Analysis

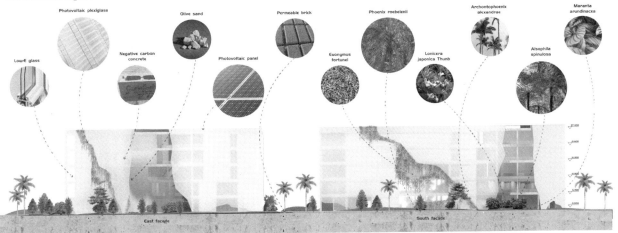

- Low-E glass
- Photovoltaic plexiglass
- Negative carbon concrete
- Olive sand
- Photovoltaic panel
- Permeable brick
- Phoenix roebelenii
- Euonymus fortunei
- Lonicera japonica Thunb
- Archontophoenix alexandrae
- Alsophila spinulosa
- Maranta arundinacea

East facade | South facade

综合奖·入围奖·零碳设计项目
Comprehensive Awards · Nomination · Zero-Carbon Design Project

注册号：101316
Register Number：101316

项目名称：风满楼
Entry Title：Wind Traveling

作者：江芊雨、戚雨田
Authors：Qianyu Jiang and Yutian Qi

作者单位：中国矿业大学
Authors from：China University of Mining and Technology

指导教师：段忠诚、邵泽彪、马全明
Tutors：Zhongcheng Duan, Zebiao Shao and Quanming Ma

指导教师单位：中国矿业大学
Tutors from：China University of Mining and Technology

■ LOCATION ANALYSIS

■ CLIMATIC ANALYSIS

■ DESIGN SPECIFICATION

The design site of this science and Technology museum is located in Guangzhou, which is a typical hot summer and warm winter climate, so cooling measures were emphasized in the design process. The design of the science and technology museum extracts the shape of the surrounding lake and hillside, simulates the flowing roof, and improves the indoor ventilation efficiency. At the same time, the science and Technology Museum adopts a large number of green building methods, fully considering the local wind and light environment, and targeted the design of passive means such as the bottom shelf, the courtyard and the atrium pulling wind, the form of self-shading, and the eaves roof. In addition, for the existing wind energy and light energy in the site, the design process considers the possibility of using a variety of different energy sources at the same time as possible, and adopts a series of active design strategies, such as photovoltaic glass, building photovoltaic integration concept, wind power generation, etc.

本文科技馆设计场地位于广州，该地属于典型的夏热冬暖气候，因此在设计过程中因器考虑降温措施。科技馆设计吸取场馆周围边湖泊山峰的形态，模拟出流线形屋面，并且提高了室内通风效率。同时，科技馆采用大量绿色建筑手法，充分考虑到当地风环境与光环境，针对性设计了底层架空、庭院及中庭通风、形体自遮阳、出檐屋顶等被动式手法。此外，针对场地在的风能资源与光能资源，设计过程中尽可能考虑了令种不同能源同时利用的可能性，采取一系列主动式设计措施，如光伏玻璃、建筑光伏一体化概念、风力发电等。

■ BULK GENERATION

1. Considering the building property line and height limit for elevated blocks
2. Considering the dominant wind direction and the cutting of the site axis to create a through air path and canyon effect
3. Reduce the west block according to functional and shading requirements
4. Forming an overhead bottom layer, conducive to passive wind pressure ventilation

5. Shaping courtyards and gray spaces, forming courtyard ventilation and roof and form shading.
6. Shaping a southword form with self shading and gray space combined with the courtyard
7. Smooth edges and corners are more conducive to the flow and guidance of wind.
8. Software calculation and comparison of roof schemes, forming curved roof skylights, shaping atrium thermal pressure ventilation

■ HISTORICAL EVOLUTION

Arcade — Bamboo tube building — Cold Alleybuilding — Patio — Wok Ear House

■ TECHNICAL SUMMARY

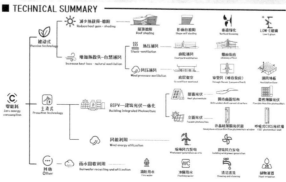

■ GENERAL PLAN 1：800

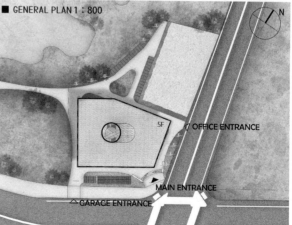

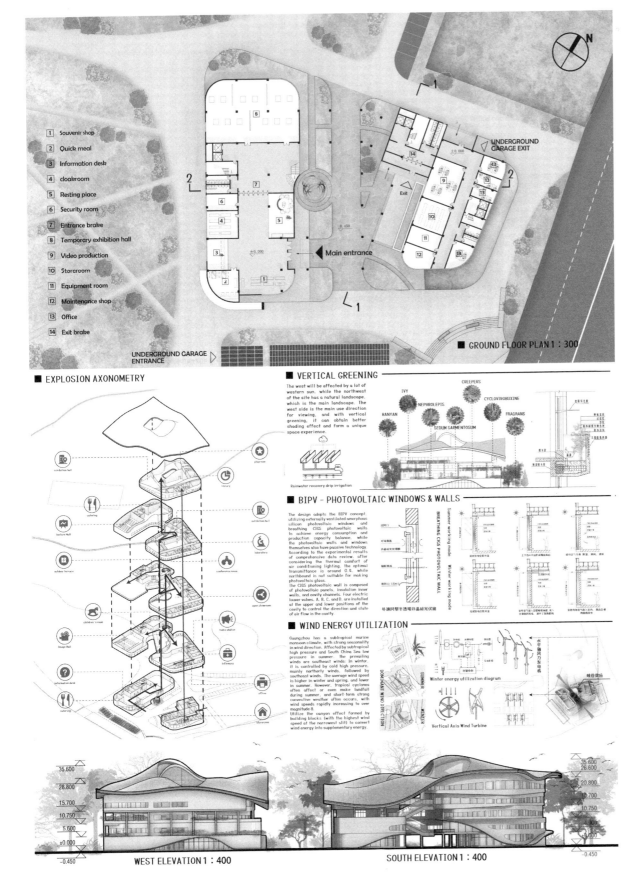

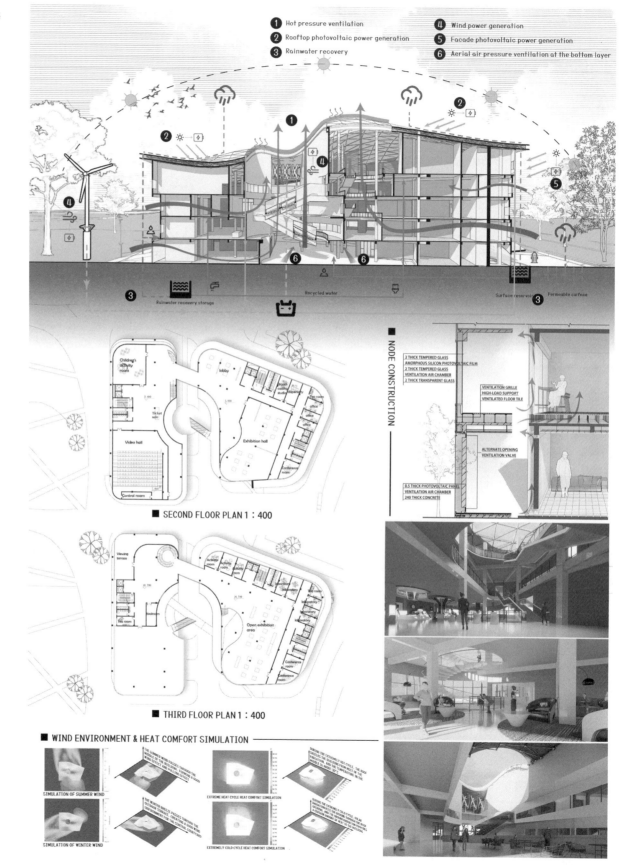

■ PLANE GRAPH

■ FOURTH FLOOR PLAN 1:400　　■ FIFTH FLOOR PLAN 1:400

■ CARBON EMISSION CALCULATION SHEET

■ PROFILING - PASSIVE APPROACH

Good ventilation and cooling are obtained from the ground floor overhead.

The temperature difference between the inside and outside of the atrium roof creates hot pressure ventilation.

The upper part of the form protrudes from the sun.

■ THERMAL ENVIRONMENT ANALYSIS

The bottom layer is ventilated and cool, and the direct part of the roof forms hot pressure ventilation.

LOW-E glass is used to effectively reduce the radiation effect.

Negative pressure is formed by the ventilation of the bottom floor, and the wind pressure difference is generated by the positive pressure of the upper part of the courtyard.

 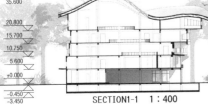 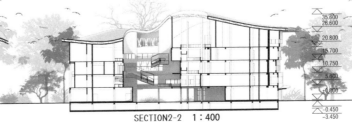

SECTION 1-1　1:400　　SECTION 2-2　1:400

综合奖・入围奖・零碳设计项目
Comprehensive Awards・Nomination・Zero-Carbon Design Project

注册号：101370
Register Number：101370

项目名称：与风同巷
Entry Title：The Evolution of Cold Alley

作者：张徐晔阳、朱珂、杨林、余嘉懿、马宏宇
Authors：Xuyeyang Zhang, Ke Zhu, Lin Yang, Jiayi Yu and Hongyu Ma

作者单位：南京工业大学、新疆大学
Authors from：Nanjing Tech University and Xinjiang University

指导教师：姜雷、胡占芳
Tutors：Lei Jiang and Zhanfang Hu

指导教师单位：南京工业大学
Tutors from：Nanjing Tech University

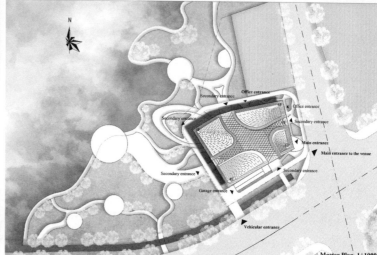

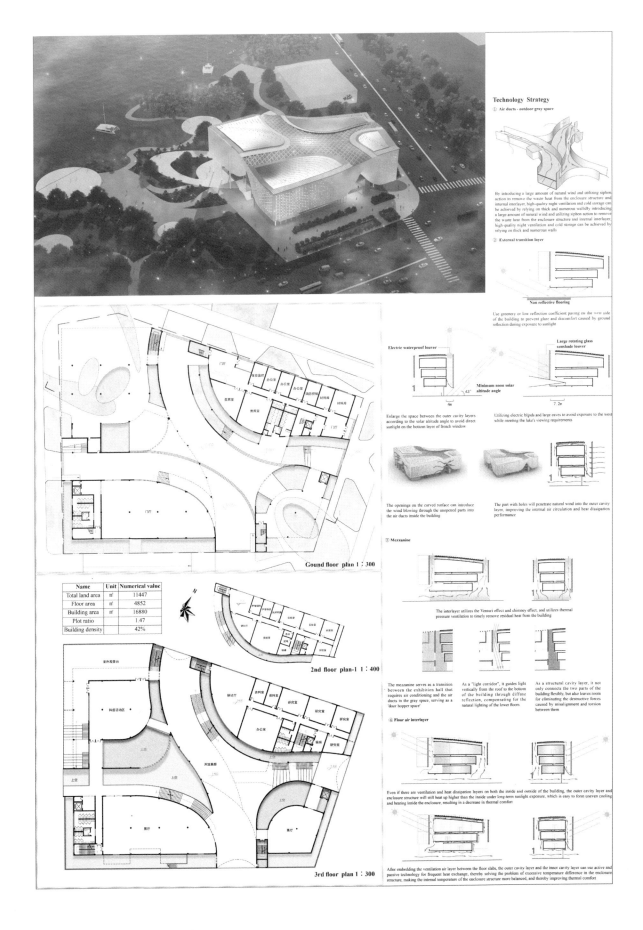

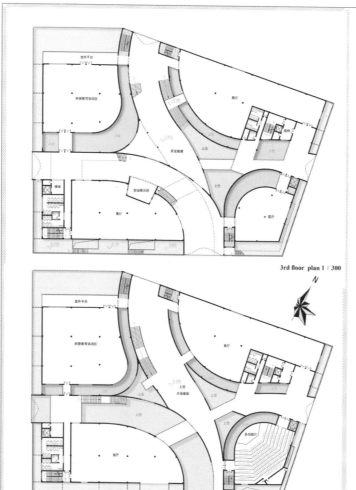

3rd floor plan 1 : 300

4th floor plan 1 : 300

View of the Sky Covered Bridge from the second-floor gallery

View of the interior of the building from the gray space of the eastern entrance

View of the open gallery from the covered bridge on the second floor

Front Elevation 1 : 600

Light Analysis

Table of Relationship between Window to Wall Ratio and Number of Floors

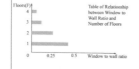

The top layer is equipped with photovoltaic skylights, which introduce light while preventing the top layer from overheating. The lower part utilizes the staggered overlap of the airducts to create a high space, which compensates for the insufficient lighting at the bottom layer

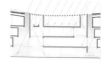

In addition, the special space formed by the inter-layer also has a certain light guiding effect due to the special structure of its inner wall. In order to reduce the temperature of the top layer and guidelight to the lower layer through reflection, the window to wall ratio of the top layer to the bottom layer increases in sequence

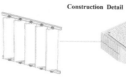

To ensure a landscape view during non Western sun exposure on the west side, a large-sized electric shutter rotary door has been introduced to solve the dual contradiction between avoiding Western sun exposure and landscape needs

Construction Detail

Due to the Venturi effect on the top of the internal interlayer, only waterproof electric louvers are needed to achieve natural thermal ventilation. However, due to the lack of thermal pressure difference on the top of the external air chamber layer, active electric fans are required to cooperate to achieve vertical ventilation

Solar battery

Low-e glass

Reinforced edging

The photovoltaic skylight on the top floor controls the intensity of skylight while also completing solar energy storage

6+20A+6+20A+6mm Tempered LOE-E insulating glass

Aluminum alloy horizontal material

Rubber cushion block

To ensure a landscape view during non Western sun exposure on the west side, a large-sized electric shutter rotary door has been introduced to solve the dual contradiction between avoiding Western sun exposure and landscape needs

Light colored fluorocarbon spraying to introduce light and withstand exposure to sunlight

Foam polyurethane insulation layer

For the convenience of transportation and construction

To ensure good cold storage, thermal insulation, and reflective performance, special thickening treatment has been applied to the walls on both sides of the interlayer inside the building

Climate Analysis

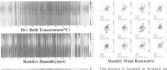

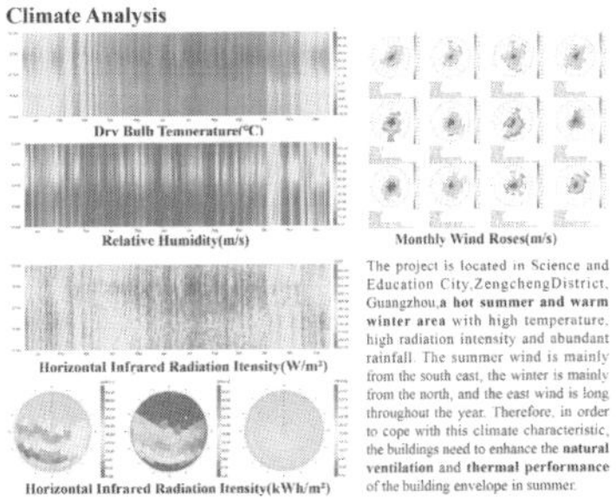

The project is located in Science and Education City, ZengchengDistrict, Guangzhou, a **hot summer and warm winter area** with high temperature, high radiation intensity and abundant rainfall. The summer wind is mainly from the south east, the winter is mainly from the north, and the east wind is long throughout the year. Therefore, in order to cope with this climate characteristic, the buildings need to enhance the **natural ventilation and thermal performance** of the building envelope in summer.

Site Problem and Measure

Planning General Plane / Current Site / Qingyun Lane / Arcade

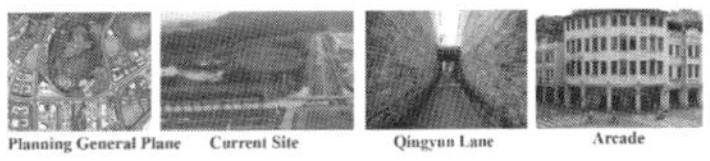

According to the general planning plan of Science and Education City and the picture of the current situation of the site, we can see that there are no tall trees and buildings around the site to shade the sun, so the building only can prevent the wall from being excessively radiated by the **self-shading method**. According to the investigation of Lingnan buildings, it is found that Qingyun lane and Arcade were used as self-shading methods in Lingnan buildings. This project takes Qingyun Lane as inspiration, and inserts a number of large-scale nonlinear **mezzanine** Spaces that run through the building blocks, and uses **Venturi effect, compression effect and lift effect** to regulate wind speed and direction.

Hierarchical Analysis

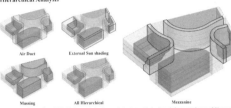

Massing / Air Duct / External Sun shading / All Hierarchical / Mezzanine

The architectural hierarchy unfolds with the ventilation ducts as the basis, transitioning through the proposed space of Qingyun Lane, centered around the massing space with the air conditioning on, and utilizing external sun shading as thermal insulation protection. The most significant aspect to explore here is **the presence or absence of transitional mezzanine levels.** Therefore, subsequent discussions on Indoor Illuminance, indoor operating temperature simulations, and other factors will predominantly focus on comparing the presence or absence of these transitional mezzanine levels.

Indoor Illuminance Analysis

Indoor Illuminance(With mezzanine space) 22 Jun 12:00

Indoor Illuminance(Without mezzanine space) 22 Jun 12:00

When we observe the two diagrams on the left, we can compare and analyze the impact of mezzanine space on the illuminance of the ground floor. From the diagrams, it is evident that without a mezzanine space, the overall illuminance decreases significantly. This decline has a severe impact on the natural daylighting of the building's air ducts and semi-outdoor walking spaces, resulting in poor lighting conditions. The presence of a mezzanine space brings a significant improvement to the lighting situation. In the case of having a mezzanine, we can see that the overall lighting conditions are relatively better. The presence of a mezzanine space provides additional **spatial levels**, which facilitates the propagation and diffusion of light, resulting in more uniform illumination throughout the space. This not only **enhances visibility** indoors but also **elevates the aesthetics and comfort** of the space.

Orientation Analysis

Optimum Orientation

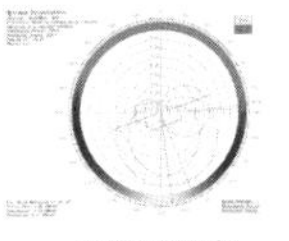

The yellow and red colors in the diagram represent the best and worst orientations, respectively, under sunlight conditions. The optimal orientation falls within the range of 150° to 210°. Taking into account **multiple factors such as summer ventilation, insulation, and the predominant southeast wind direction**, it was ultimately decided to orient the building towards 150°. This choice not only fulfills the optimal building orientation but also addresses the issue of wind channeling within the structure, ensuring effective wind flow management.

Note: All analysis diagrams on this page are not aligned with the true north-south direction. They have been rotated clockwise by 30°. The north direction is shown in the diagram on the right.

Orientation Analysis

Direct Sun Hours 22 Jun 8:00–18:00

Direct Sun Hours 22 Dec 8:00–18:00

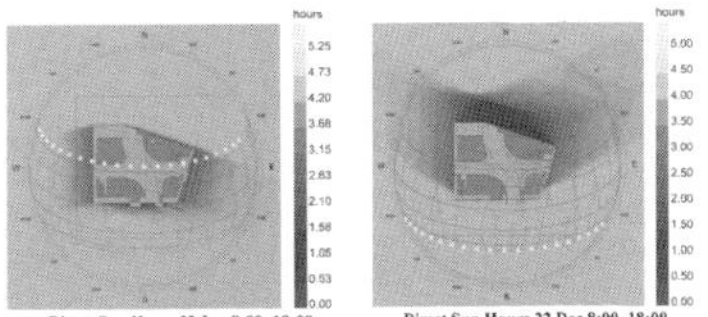

Due to the proximity of the site to the Tropic of Cancer, the building creates a smaller shadow range on the site. Therefore, it is possible to incorporate local characteristic **plants for building carbon sequestration** while meeting the functional design requirements of the site.

Direct Sun Hours 22 Jun 8:00–18:00 / Direct Sun Hours 22 Dec 8:00–18:00

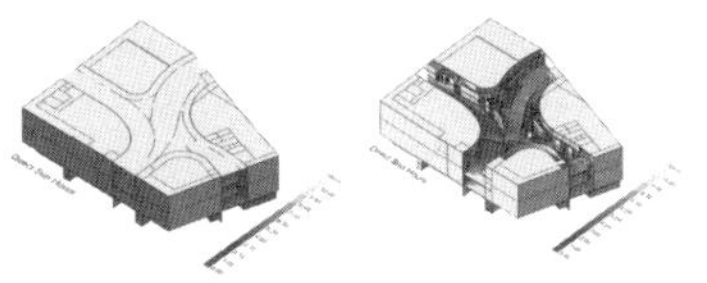

In the summer, when the roof is exposed to high levels of solar radiation, it is advisable to use skylight shading to reduce heat gain. On clear days with a sunshine duration exceeding 10 hours, it is suitable to install rea of photovoltaic solar panels on sloped roofs to harness solar energy. During the winter, when the temperature is around 10°C, it is beneficial to open skylights to allow **passive solar heating**. This allows sunlight to enter the space, contributing to natural heating and reducing the reliance on active heating systems.

Ventilation Simulation

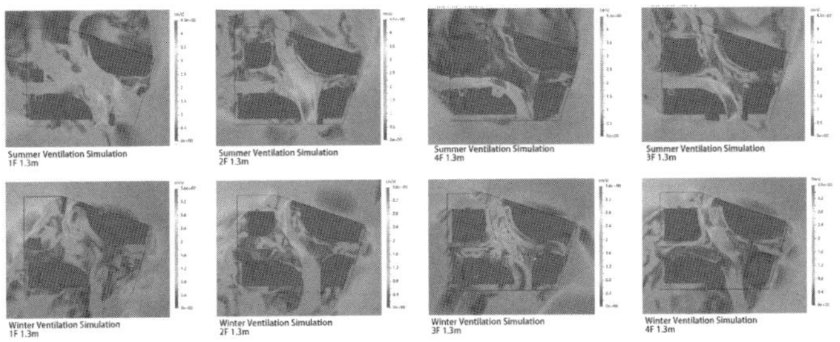

Summer Ventilation Simulation 1F 1.3m / 2F 1.3m / 3F 1.3m / 4F 1.3m

Winter Ventilation Simulation 1F 1.3m / 2F 1.3m / 3F 1.3m / 4F 1.3m

The simulation conditions of the model in summer are as follows: southeast wind 2.7m/s; Winter simulation conditions: North wind 2.3m/s. By examining the charts, we can draw important conclusions. Firstly, by designing the air duct openings to align with the prevailing wind direction, we have successfully introduced natural ventilation into the building's air ducts. This design decision provides significant potential for energy-saving and emission reduction. During summer, the airflow within the air ducts effectively removes heat, lowers indoor temperatures, and reduces the burden on the air conditioning system. This helps conserve energy and decrease carbon emissions and enhances comfort and indoor air quality.

Passive Strategy Analysis

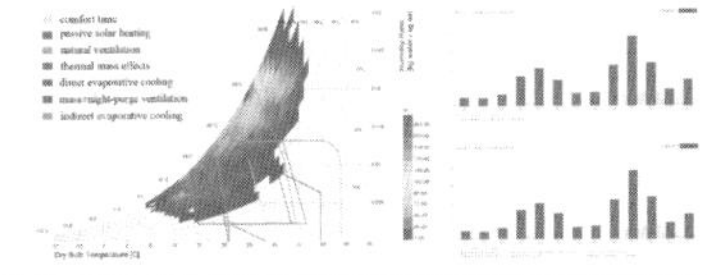

Based on a passive-first strategy, six different passive techniques have been added to the psychrometric chart. By comparing the illustrations, two most effective passive energy-saving techniques are identified: **natural ventilation and insulated walls with night ventilation**. Additionally, the illustrations indicate that even with the inclusion of passive techniques, a majority of the hours experience temperatures that do not meet the comfort requirements, requiring HVAC **intervention** to ensure human comfort.

Indoor Operature Temperature Simulation

Operating Temperature(With mezzanine space) 22 Jun 17:00 / 22 Jun 12:00

Operating Temperature(Without mezzanine space) 22 Jun 17:00 / 22 Jun 12:00

By comparing the visual images of indoor operating temperature in summer with or without mezzanine space, it can be seen that the volume with mezzanine space has a lower temperature than the volume without mezzanine space, and has a better cooling effect.

From the operating temperature diagram at 12 noon in summer on the fourth floor, it is obviously found that heat accumulation occurs in the center of the space without mezzanine, which is extremely unfavorable for the progress of architectural design. However, after the addition of mezzanine, this phenomenon obviously disappears, which fully proves the thermal insulation and cooling effect of the mezzanine space.

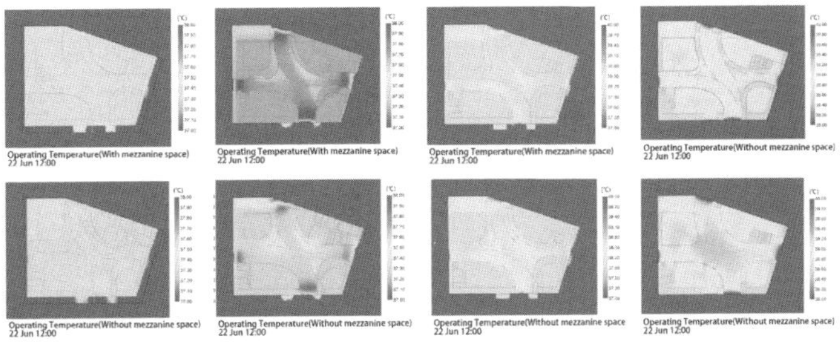

Zone Operative Temperature and Comfort

According to the ISO 7730-2005 thermal comfort standard, the recommended range for the PMV (Predicted Mean Vote) index is -0.5 to +0.5. The illustration demonstrates that a certain level of comfort is achieved solely through natural ventilation and the effect of intermediate spaces, highlighting the rationality of air duct orientation and space configuration.

Building Intelligent Control System

In this project, it is worth mentioning some intelligent green-saving technologies, such as the intelligent control system of air volume according to the seasonal climate change, the optical storage direct and flexible integrated system that can store electricity according to the intelligent power consumption.

Analysis of Active and Passive Energy Saving Techniques

1. Photovoltaic solar panel power generation system: The use of renewable sources of light to provide electricity for the building.

2. Ventilation fan + intelligent louver: Through the participation of mechanical ventilation to enhance the effect of wind pressure and wind direction.

3. Water collection and storage system: Rainwater is collected and stored in building water supply system through an organized drainage system for toilet water...

4. Photovoltaic cell power generation system: photovoltaic cell boxes are arranged on the triangular skylight to maximize the use of the radiant solar energy on the roof.

5. Tunnel air system: The use of soil specific heat capacity is large, warm in winter and cool in summer to achieve energy saving and emission reduction.

6. Polyurethane phase change insulation wall: latent heat is formed through polyurethane phase change characteristics and then eliminated by night ventilation.

7. Polyurethane phase change insulation wall: latent heat is formed through polyurethane phase change characteristics and then eliminated by night ventilation.

8. Three-glass two-cavity glass curtain wall: It has one glass and one cavity more than the common curtain wall, which improves the air tightness and thermal insulation performance of the curtain wall envelope.

9. Electric louver: Intelligent louver control system is added to meet the requirements of sunshine and west sun protection.

Calculation of Building Carbon Emissions

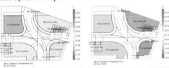

Calculation Results of Carbon Emissions during the Operational Phase of Buildings			
Project		Carbon Emissions kgCO₂	Area ㎡	Carbon emissions per unit area kgCO₂·㎡
Carbon Emission Term	HVAC			
	Domestic Hot Water			
	Illumination			
	Elevator			
Carbon Reduction Term	Renewable Energy			
	Building Carbon Sink			
	Passive Techniques			
Sum				

综合奖・入围奖・零碳设计项目
Comprehensive Awards · Nomination · Zero-Carbon Design Project

注册号：101371
Register Number：101371

项目名称：入风吟
Entry Title：Chant in Wind

作者：黄晓、张黔渝、浮英媛
Authors：Xiao Huang, Qianyu Zhang and Yingyuan Fu

作者单位：昆明理工大学
Authors from：Kunming University of Science and Technology

指导教师：谭良斌
Tutor：Liangbin Tan

指导教师单位：昆明理工大学
Tutor from：Kunming University of Science and Technology

入风吟

项目位于广州增城区，受当地气候和文化的影响，设计希望建筑是开放的，给当地居民提供自由交流的活动空间。

建筑通过抬高地层面、架空、设置半室外平台等形式创造了通透但有遮蔽功能的市民空间。设计从广州传统民居竹筒屋的通风原理中得到启示，在建筑中设置风道和小天井共用来通风。同时，利用多功能表皮进一步回应当地气候，集合了太阳能板发电、垂直通风、智能遮阳、光线柔化等功能，除此之外，还进行了光伏屋顶、生态水池、屋顶花园等主、被动式设计。

■ Design Description

The project is located in Zengcheng District, Guangzhou, and is influenced by the local climate and culture. We hope that the building is open and provides free space for local residents to exchange activities. The building creates a transparent but sheltered civic space through elevated levels, elevated structures, and the installation of semi outdoor platforms.

The design draws inspiration from the ventilation principles of traditional bamboo tube houses in Guangzhou, where air ducts and small courtyards are installed for ventilation. At the same time, multifunctional skins are used to further respond to the local climate, integrating functions such as solar panel power generation, vertical ventilation, intelligent shading, and light softening. In addition, active and passive designs such as photovoltaic roofs, ecological pools, and rooftop gardens have also been carried out.

■ Climate Analysis

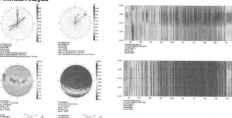

Guangzhou belongs to a warm winter and hot summer area, with high temperatures and humidity in summer, and strong solar radiation in Guangzhou. Therefore, buildings mainly consider the requirements of ventilation, shading, and insulation, and design active and passive systems to fully utilize solar energy resources.

■ Design Ideas and Strategies

Site Analysis

About site
The site is located on the central axis of Guangzhou Science and Education City and in the comprehensive functional area. The north side of the site is relatively high, while the west side is relatively low.

About road
There are roads on the south and east sides of the site, but the road on the south side is wider and the main pedestrian flow is towards the south. Therefore, the main entrance is set on the south side.

About building
There are few buildings around the existing site, and a 110kV substation is planned to be built on the southeast side of the site.

About Water
The west side of the site is a water surface, and the site is separated from the south road by a water channel. An ecological water pool can be set up in conjunction with the water channel.

About Plant
The surrounding green environment of the site is good, with rich vegetation. There is a long irregular space on the north side of the site, which combines the water surface and surrounding plants, sets up a platform and greenery.

Concept Extraction

Inspiration is drawn from the ventilation form of traditional bamboo tube houses in Guangzhou, which utilizes halls, cold alleys, and courtyards, and is applied to the ventilation of this building.

Site red line

Form and Concept Generation

Phrase 1 Determine the floor height based on the construction site

Phrase 2 Placing into cold alleys to divide building blocks

Phrase 3 Combining multiple small courtyards in cold alleys

Phrase 4 Adjust the size of the block according to the building function

Phrase 5 Determine the building blocks and wind corridors

Continuously adjust **the width of the wind corridor** according to the simulation to achieve the best results

Phrase 6 Determine building height based on the direction of summer winds

Phrase 7 Elevate the ground level, create entrances and activity squares

Phrase 8 Elevated space to enhance spatial openness

Phrase 9 Set aside platform and gray space

Phrase 10 Set up the sunshade and connect the various blocks

Phrase 11 Set up building skins and incorporate architectural design details

General Layout

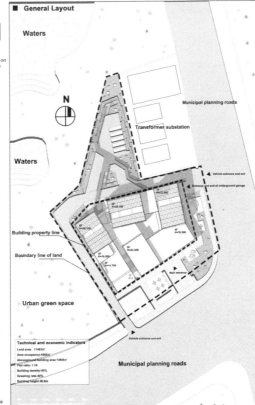

Technical and economic indicators
Land area: 11463m²
Area occupancy: 4560m²
Aboveground building area: 13659m²
Plot ratio: 1.19
Building density: 40%
Greening rate: 40%
Building height: 36.8m

Section View

1. Hallway
2. Material room
3. Exhibition Production Room
4. Exhibition storage room
5. Channels
6. Production and maintenance workshop
7. Walkways
8. Technical Archives
9. Exhibition Hall 1
10. Exhibition Hall 2
11. Exhibition Hall 3
12. Reading Area
13. Office 1
14. Office 2
15. Office 3
16. Office 4
17. Walkways
18. Lounge
19. Equipment layer

Section A-A

1. Production and maintenance workshop
2. Walkway
3. Children's Exhibition Hall
4. Public activity areas
5. Exhibition Hall 1
6. Souvenir Store
7. Aisle
8. Exhibition Hall 2
9. Waiting and Rest Area
10. Ball Screen Cinema
11. Interactive demonstration area 1
12. Cafe

Section B-B

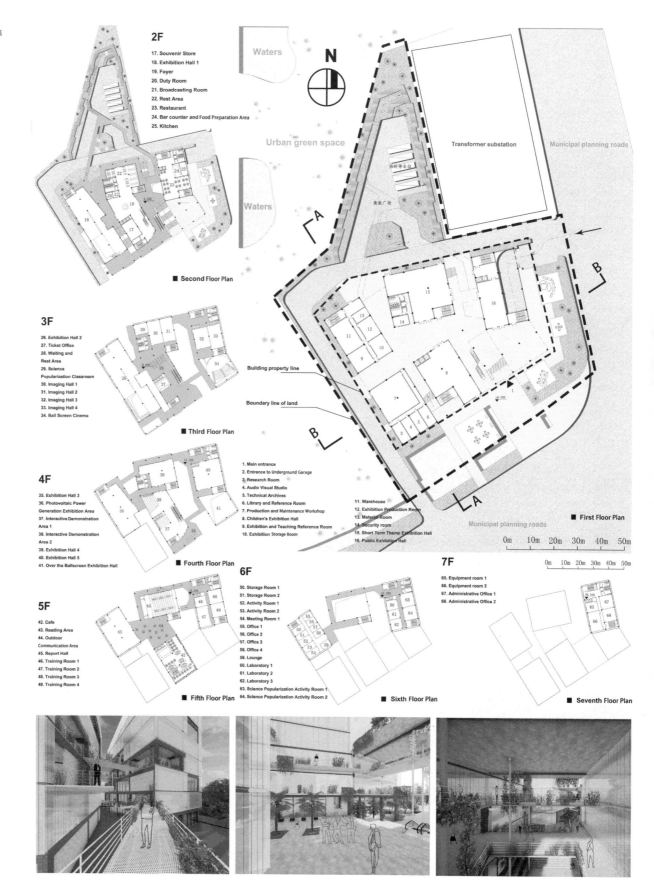

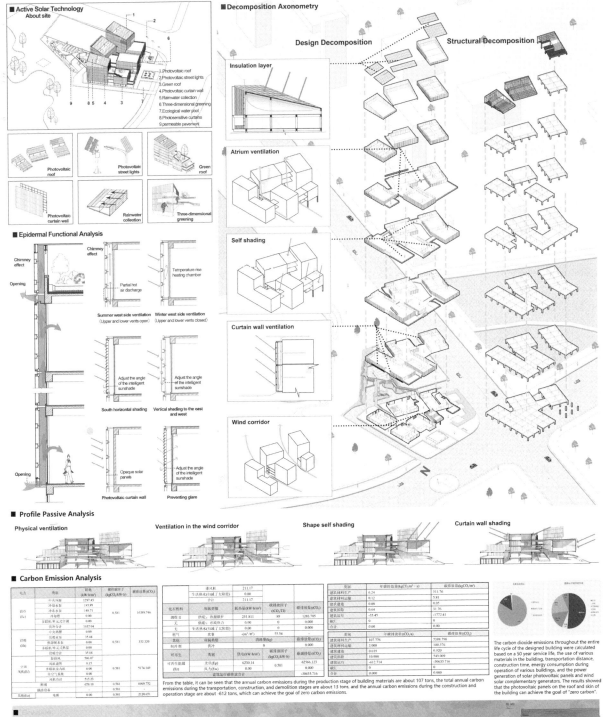

综合奖·入围奖·零碳设计项目
Comprehensive Awards · Nomination · Zero-Carbon Design Project

注册号：101392
Register Number: 101392

项目名称：光檐风堂
Entry Title: Wind Hall under Solar Photovoltaics

作者：谢楚湉、慈青松、黄丽文
Authors: Chutian Xie, Qingsong Ci and Liwen Huang

作者单位：北京建筑大学、南京工业大学
Authors from: Beijing University of Civil Engineering and Architecture, Nanjing Tech University

指导教师：孙立娜、孙嘉男
Tutors: Lina Sun and Jianan Sun

指导老师单位：北京建筑大学、北京优优星球教育科技有限公司
Tutors from: Beijing University of Civil Engineering and Architecture, Beijing Yoyo Star Education and Technology Co., Ltd.

■ Design Specification

科技文化馆采用整体式设计，为钢筋混凝土结构，利用板片穿插与完整体块形成对比，一层采用无气候边界的开放式布局，从二层向上为封闭平层，作为主要功能区域。建筑中央形成内庭井垂直贯通建筑，形成拔风结构提升室内通风效果。玻璃幕墙全部采用光伏玻璃，屋顶板铺设太阳能板，达到光-储-直-柔。独特的大板结构在屋面放大通风瓦结构，梳理屋顶风环境；在建筑下形成灰空间增强建筑自遮阳，减少室内太阳直射；在功能上让建筑形态更为活泼自由。

The overall science and technology cultural center adopts a monolithic design, which is a reinforced concrete structure, and the use of slabs interspersed to contrast with the complete block. The ground floor adopts an open layout without climatic boundaries, and the closed flat floor is the main functional area from the second floor upwards. The center of the building forms an inner courtyard and runs vertically through the building, forming a pull-out structure to improve the indoor ventilation effect. The glass curtain wall is all made of photovoltaic glass, and the roof panels are laid with solar panels to achieve light-storage-straight-soft. The unique large-plate structure enlarges the ventilation tile structure on the roof to comb the roof wind environment; The formation of gray space under the building enhances the self-shading of the building and reduces direct indoor sunlight; In terms of function, the architectural form is more lively and free.

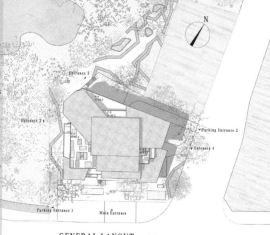

GENERAL LAYOUT 1:500

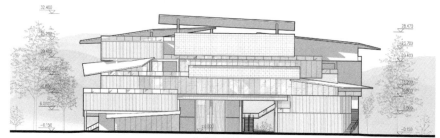

SOUTH ELEVATION 1:300

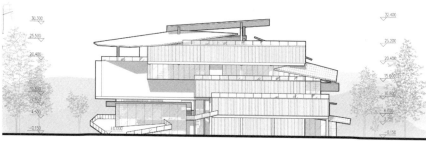

WEST ELEVATION 1:300

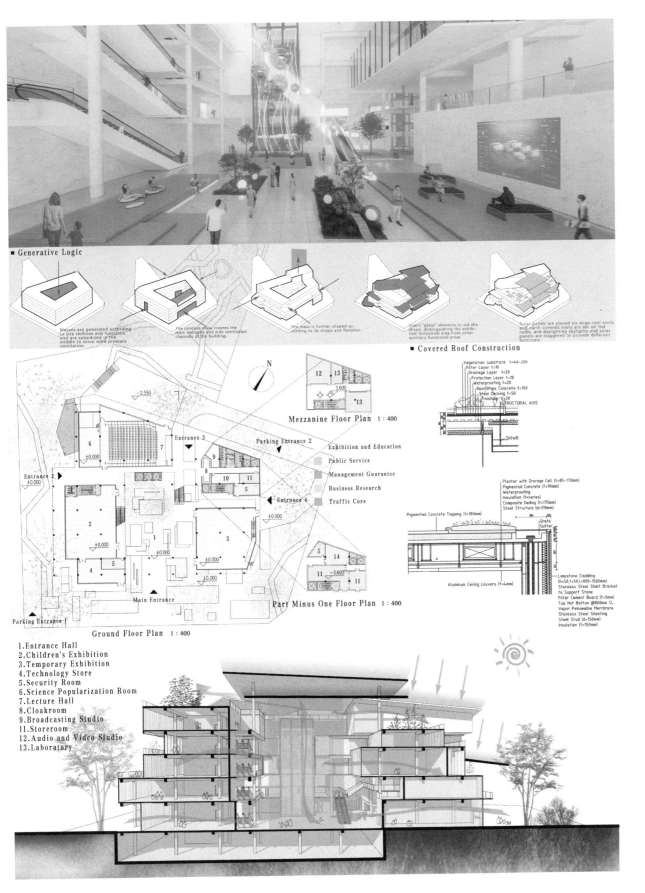

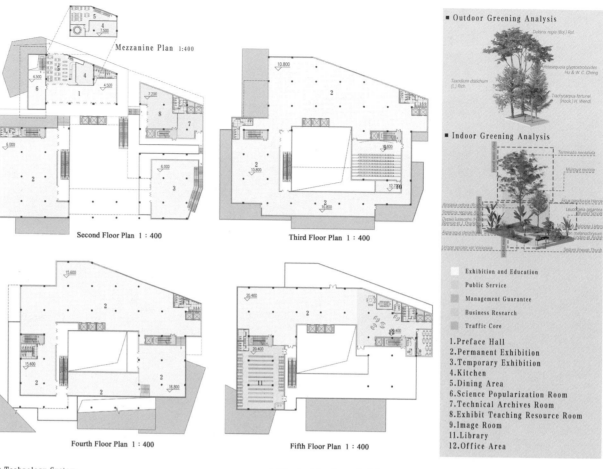

Standard for Green Building

Energy Consumption Comparison				
Type of energy consumption	Designed building		Reference building	
Heating set (kW·h)	E	48751.55	E	38025.61
Air-conditioner set (kW·h)	E	327307.6	E	463672.8
Lighting (kW·h)	E	262448.32	E	294996.85
All-year energy consumption (kW·h)	B	638507.47	B	796695.26
Reduction of energy consumption		19.86%		

	Total annual radiation(kW·h/m²)	200
solar energy	PV module installation area(m²)	1000
	Overall efficiency factor(0~1)	0.3
	Annual photovoltaic power generation(kW·h/a)	60000

Category	Design the annual operating carbon emissions per unit area of the building[kgCO₂/(m²·a)]	Refer to the annual operating carbon emissions per unit area of the building[kgCO₂/(m²·a)]	Optimize the proportion%	Reduction of annual operating carbon emissions per unit area [kgCO₂/(m²·a)]
heating	2.3	1.89	-22.09	-0.42
Air-conditioner	15.46	23	32.78	7.54
Lighting	12.4	14.63	15.28	2.24
Renewable energy	2.83	0	—	—
Total	27.33	39.52	30.85	12.19

renewable energy Comparison			
Designed building		Reference building	
E	60000	E	0
Reduction of annual operating carbon emissions per unit area[kgCO₂/(m²·a)]			12.19

■ Technology System

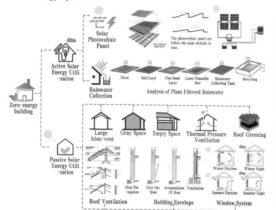

■ Building Ventilation

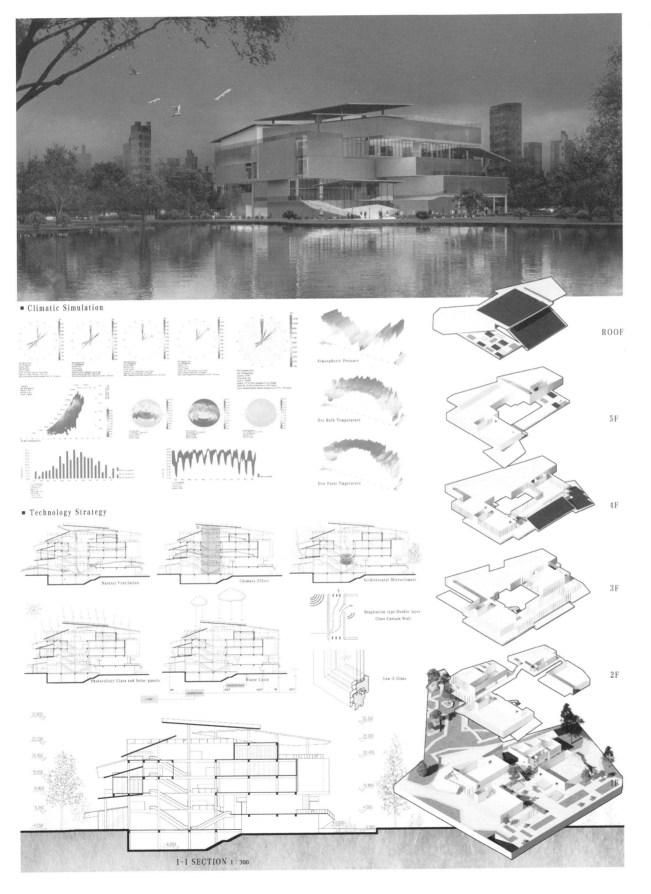

综合奖・入围奖・零碳设计项目
Comprehensive Awards・Nomination・Zero-Carbon Design Project

注册号：101396
Register Number：101396

项目名称：星落
Entry Title：Meteor

作者：蔡秀蕴、刘贤、万子豪、黄伟峰、高昕冉、蒋依珊
Authors：Xiuyun Cai, Xian Liu, Zihao Wan, Weifeng Huang, Xinran Gao and Yishan Jiang

作者单位：浙江理工大学
Authors from：Zhejiang Institute of Science and Technology

指导教师：文强
Tutor：Qiang Wen

指导教师单位：浙江理工大学
Tutor from：Zhejiang Institute of Science and Technology

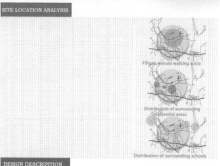
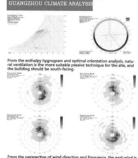

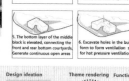

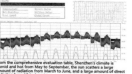
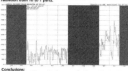

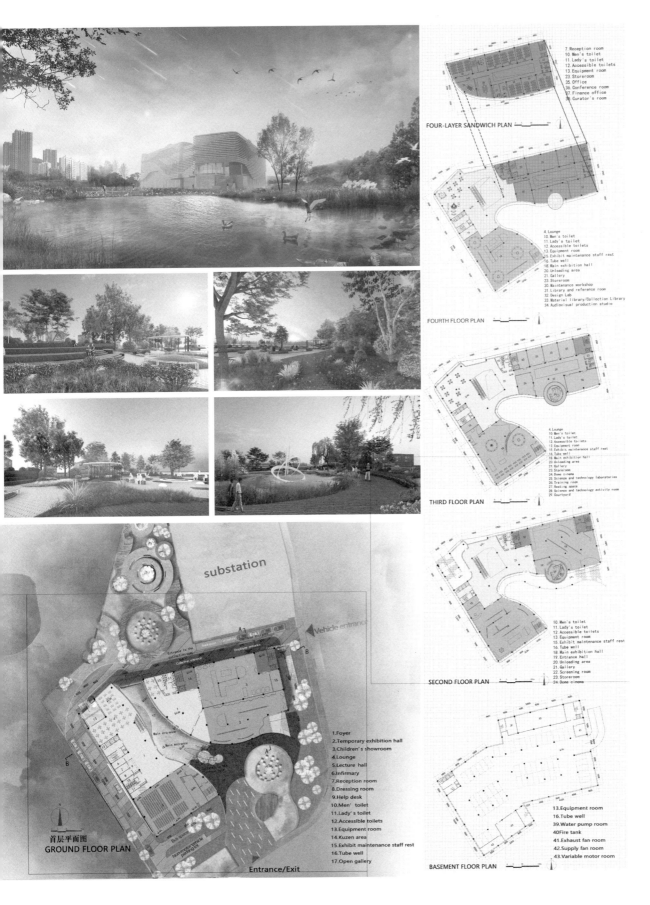

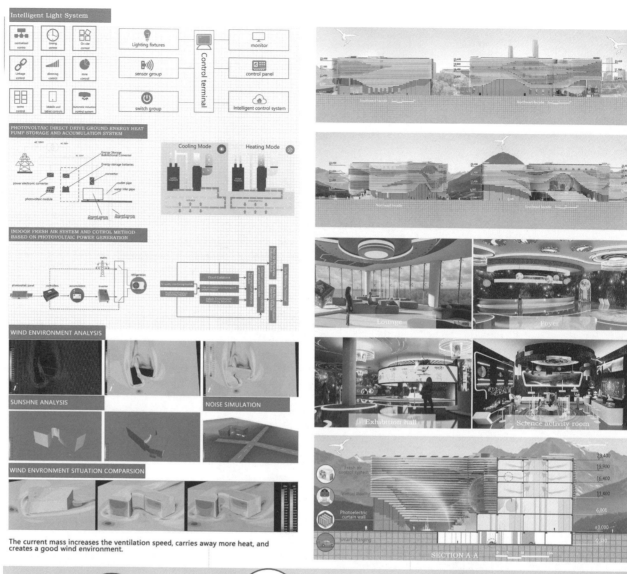
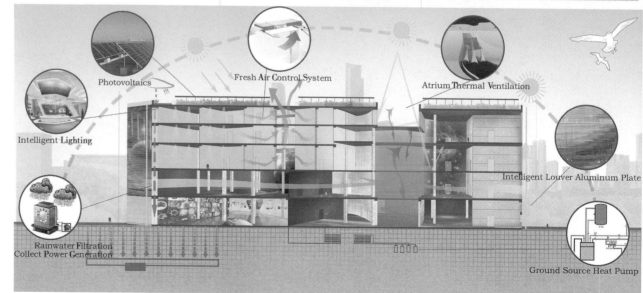

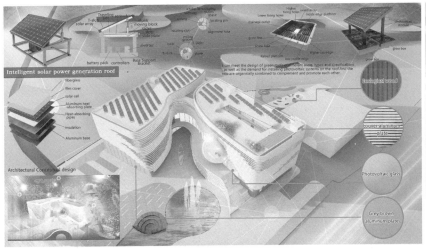

Intelligent solar power generation roof

Architectural Conceptual design

Starlight Dreamland Park

Meteor Plaza

Star River Exploration

OPTICAL STORAGE DIRECT FLEXIBLE TECHNOLOGY

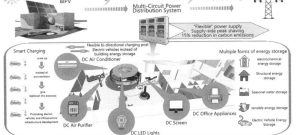

COMPUTATIONAL ANALISIS

Calculation results of renewable energy in buildings: 91%

Conclusion of energy-saving design regulatory inspection

Explore the Planet

DIAGRAM OF DESIGN PROCESS

FULLY OPENABLE DOUBLE-SKIN BREATHABLE ENERGY -SAVING GLASS CURTAIN WALL SYSTEM

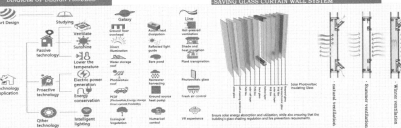

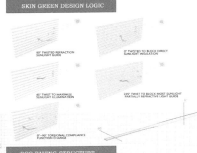
Meet the Stars

PHOTOVOLTAIC POWER GENERATION

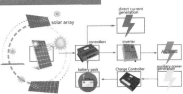

By changing the angle of the solar arrays, the solar arrays are oriented towards the sun at an optimal angle to fully utilize the solar energy.

RAINWATER POWER GENERATION DEVICE

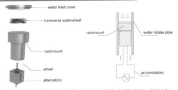

The rainwater is stored in the reservoir and flows into the pressurized header of the inlet pipe, which has the function of pressurizing the water flow and generating a vortex. The pressurized pipe head has the function of pressurizing water flow, and together with the transverse water wheel, the faster water flow makes the transverse water wheel rotate faster, and then connects to the generator, which makes the water in the water inlet pipe. The connection to the generator enables the generator to work stably even when the water pressure in the water inlet pipe is low.

SKIN GREEN DESIGN LOGIC

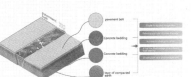

ECO PAVING STRUCTURE

PASSIVE DESIGN ANALYSIS

- Building Ventilation - Application of Green Passive Technology
- Building Daylighting - Passive Solar Energy Design
- The principle of hot pressure ventilation forms wind vortices to accelerate air circulation and achieve building cooling.

- Square pool reduces prevailing hot air temperature
- Building U-shaped guiding wind direction
- 1. Added local areas within the courtyard Ventilating vortex and pulling out wind to form an air duct
- 2. Atrium: Change the roughness of the underlying surface to form local areas. Microcirculation of air convection and optimization of wind environment

综合奖・入围奖・零碳设计项目
Comprehensive Awards · Nomination · Zero-Carbon Design Project

注册号：101398
Register Number：101398

项目名称：生长・光谷
Entry Title：Growth, Optical Valley

作者：王一琳、武昊阅、杨沛东、张琦
Authors：Yilin Wang, Haoyue Wu, Peidong Yang and Qi Zhang

作者单位：南京工业大学
Authors from：Nanjing Tech University

指导教师：董凌
Tutor：Ling Dong

指导教师单位：南京工业大学
Tutor from：Nanjing Tech University

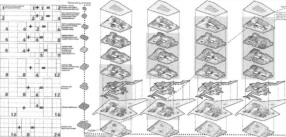

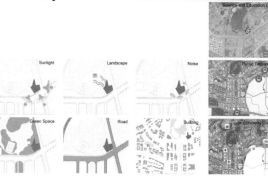

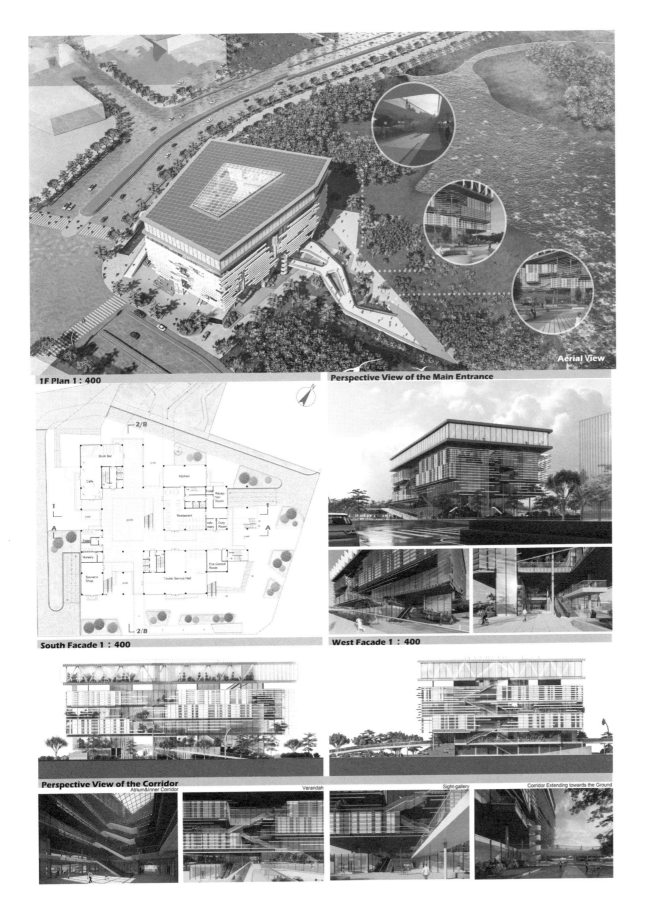

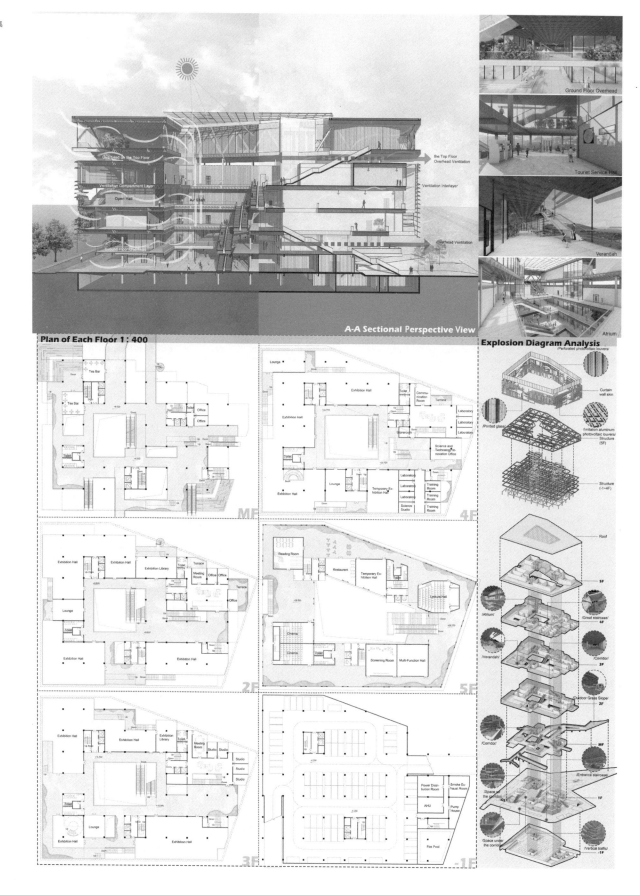

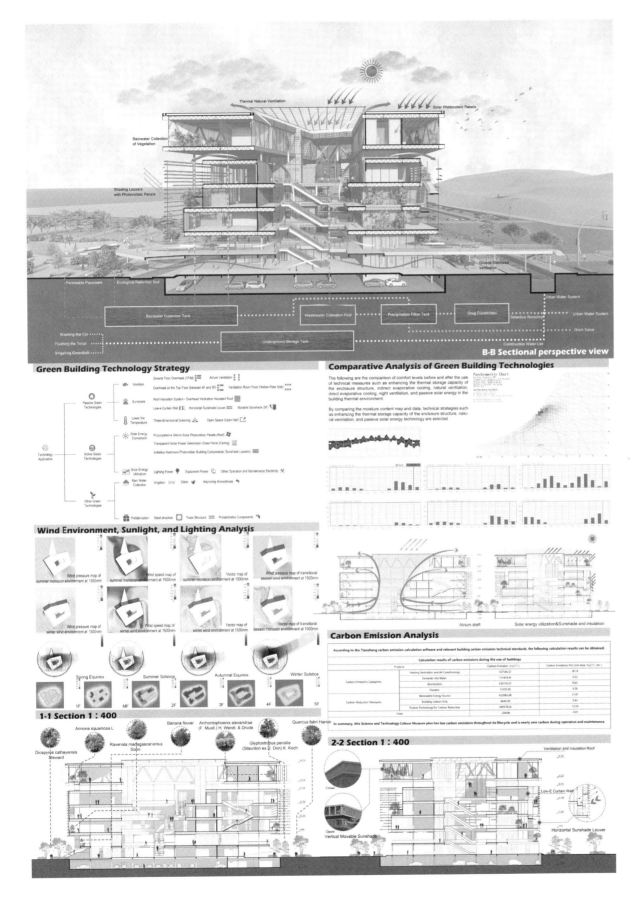

综合奖·入围奖·零碳设计项目
Comprehensive Awards · Nomination · Zero-Carbon Design Project

注册号：101403
Register Number：101403

项目名称：风巷赏，光塔下思
Entry Title：Enjoying in Windy Alley, Thinking under the Tower of Light

作者：杨静怡、郑晨雨、俞俊琛、汤宇豪
Authors：Jingyi Yang, Chenyu Zheng, Junchen Yu and Yuhao Tang

作者单位：福州大学
Authors from：Fuzhou University

指导教师：吴木生
Tutor：Musheng Wu

指导教师单位：福州大学
Tutor from：Fuzhou University

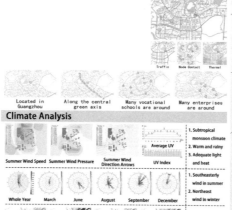

建筑方案设计说明

本方案以被动式技术为设计出发点，结合岭南传统民居的通风隔热手法与功能层次关系，以及广州科教城的人群需求，赋予科技馆"绿色、低碳、节能"集成与"教、产、展"一体化的属性。通过场地的水体绿化布置缓冲热量，顺应夏季风向设置风道破解体块，利用太阳能烟囱与灰空间加强自然通风与隔热效果。设置太阳能光伏系统以及雨水收集系统，进行屋顶一体化设计与场地设计，实现可持续能源的回收利用与建筑产能。

Design Description

This scheme takes passive technology as the design starting point, combines the ventilation and heat insulation methods and functional level relationship of traditional houses in Lingnan, and the needs of the population of Guangzhou Science and Education City, and gives the science and technology museum the attributes of green, low-carbon, energy-saving integration and education, production and exhibition integration. The water and greenery of the site are arranged to buffer the heat, the air duct is set up to crack the body block according to the summer wind direction, and the natural ventilation and heat insulation effect are strengthened by using the solar chimney and gray space. Solar photovoltaic systems and rainwater harvesting systems are installed, and integrated roof and site design are carried out to achieve sustainable energy recovery and building capacity.

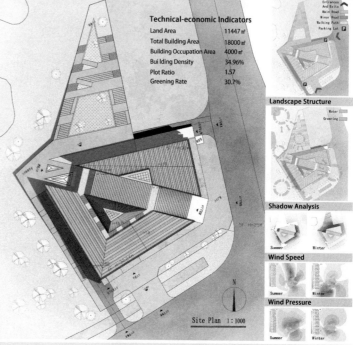

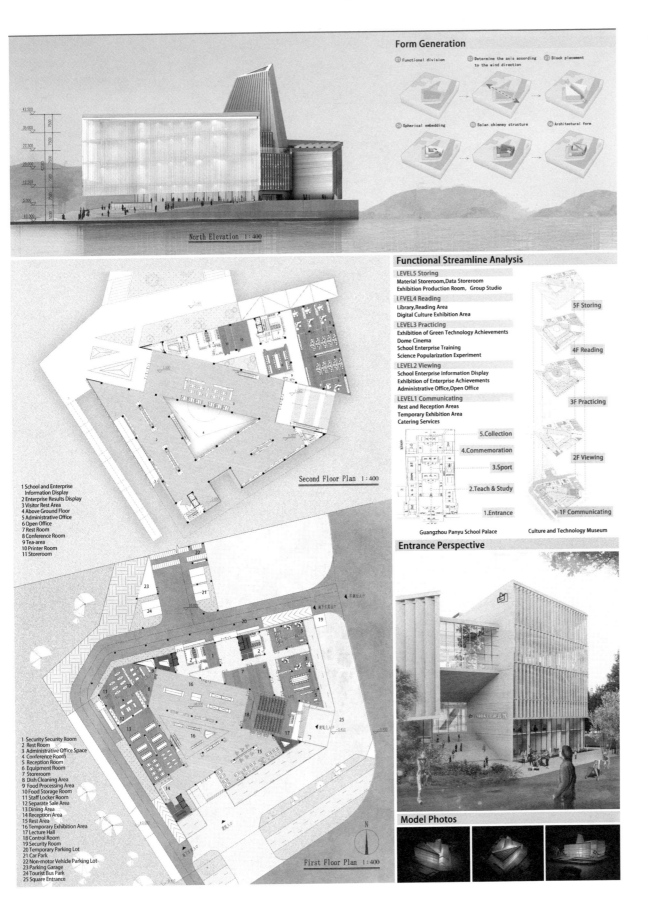

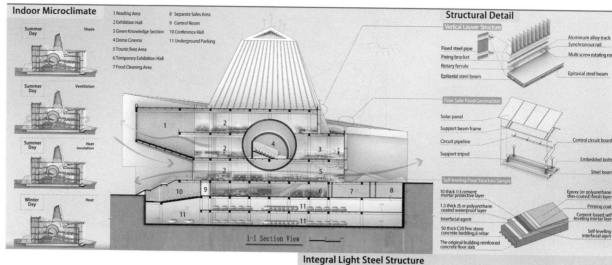
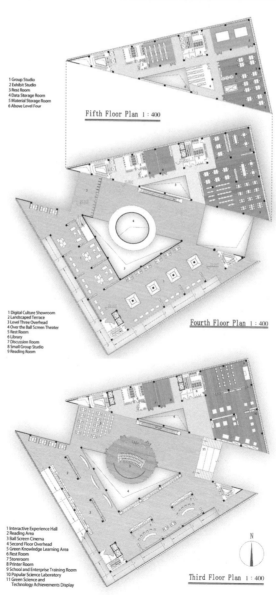
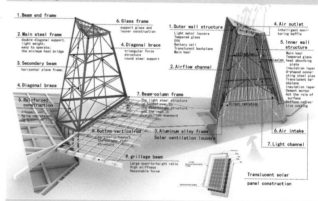

Zero Energy Building Technical Indicators

Zero Carbon Building Technical Indicators

Structural Explosion Diagram

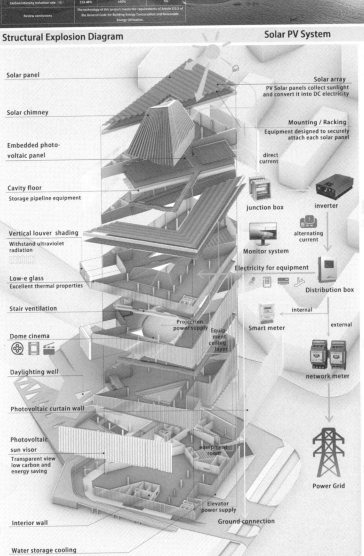

- Solar panel
- Solar chimney
- Embedded photo-voltaic panel
- Cavity floor
  Storage pipeline equipment
- Vertical louver shading
  Withstand ultraviolet radiation
- Low-e glass
  Excellent thermal properties
- Stair ventilation
- Dome cinema
- Daylighting well
- Photovoltaic curtain wall
- Photovoltaic sun visor
  Transparent view low carbon and energy saving
- Interior wall
- Water storage cooling

Solar PV System

- Solar array: PV Solar panels collect sunlight and convert it into DC electricity
- Mounting / Racking: Equipment designed to securely attach each solar panel
- direct current
- junction box
- inverter
- alternating current
- Monitor system
- Electricity for equipment
- Distribution box
- internal
- Smart meter
- external
- network meter
- Power Grid

SC-EAHE Coupled Batural Ventilation System

Pipes are buried underground and equipment for heat exchange with the soil extracts the heat energy of the soil into the air inside the pipes to cool the building in summer. At the same time, the solar chimney is used to heat press to strengthen the natural ventilation head and air volume.

Rainwater Utilization

01 Rainwater Harvesting

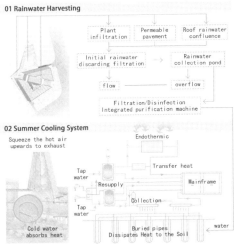

02 Summer Cooling System

03 Water Recycling and Reuse

Sprinkler Irrigation | Equipment Cooling | Pavement Washing

综合奖·入围奖·零碳设计项目
Comprehensive Awards · Nomination · Zero-Carbon Design Project

注册号：101451
Register Number：101451

项目名称：羊城叠翠
Entry Title：Yangcheng Ecology

作者：周凯文、郑泽楠、金函羽、肖雨昕
Authors：Kaiwen Zhou, Zenan Zheng, Hanyu Jin and Yuxin Xiao

作者单位：福州大学
Authors from：Fuzhou University

指导教师：王炜
Tutor：Wei Wang

指导教师单位：福州大学
Tutor from：Fuzhou University

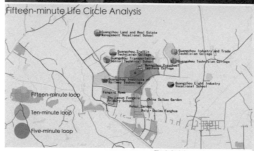

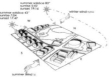

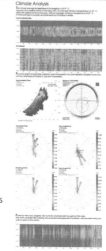

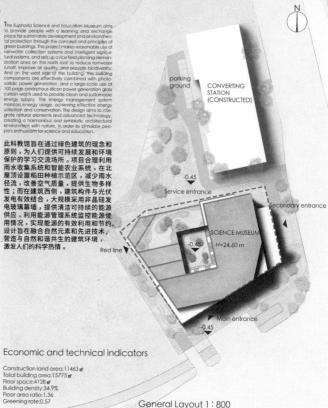

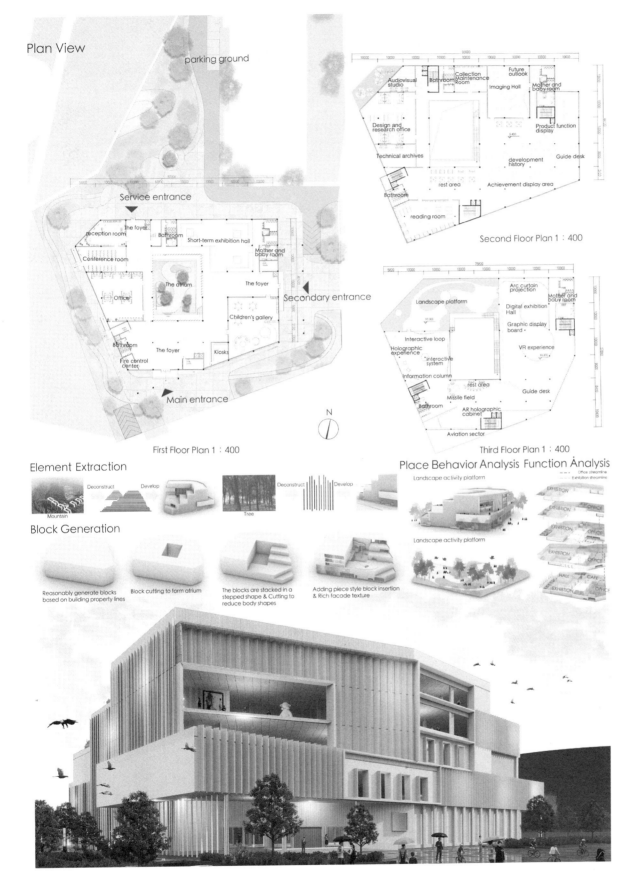

Fourth Floor Plan 1:400

East Facade 1:400

East Facade 1:400

Fifth Floor Plan 1:400

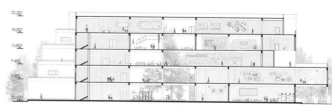
1-1 Profiles 1:400

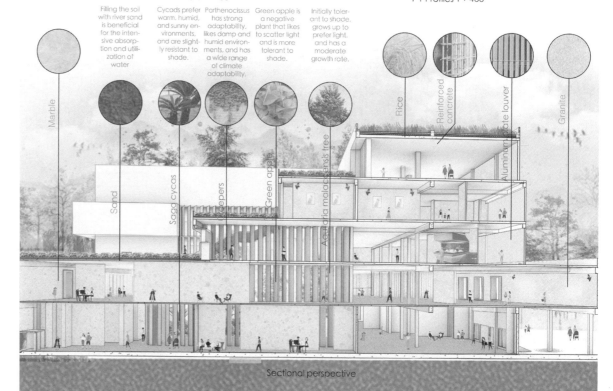
Sectional perspective

Energy-saving Analysis

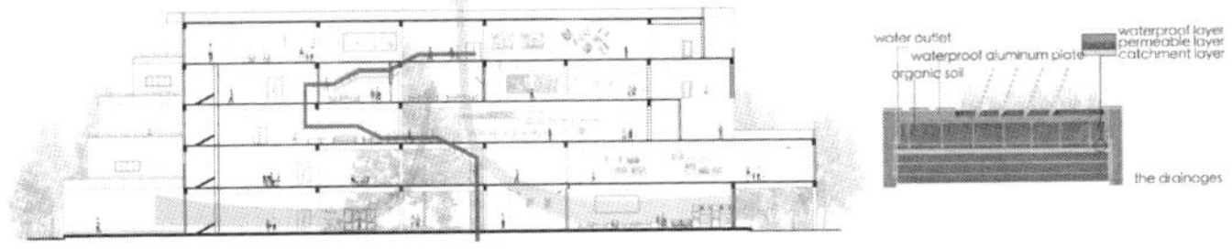

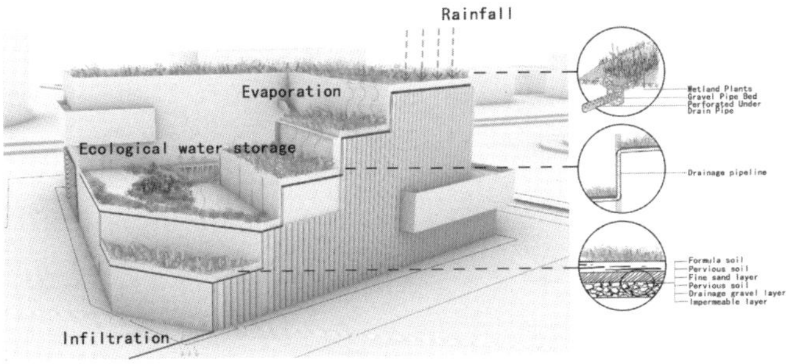

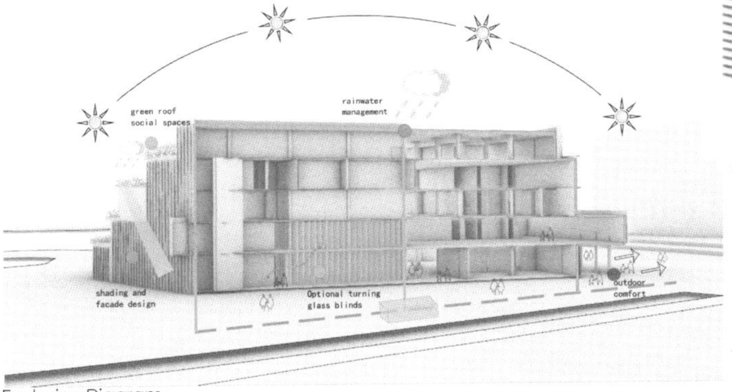

Specific Component Analysis

Photovoltaic curtain wall is a green building project that combines solar power generation systems with curtain wall engineering. Photovoltaic curtain wall can not only meet the design requirements of glass curtain wall lighting and insulation, but also meet the functions of photovoltaic power generation, maximizing the use of natural energy, and providing a green and environmentally friendly life.

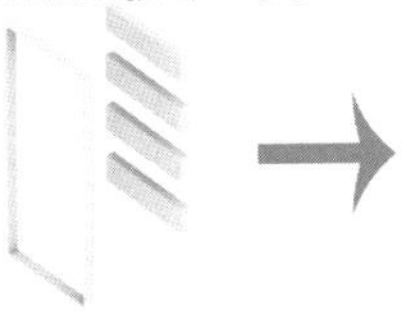

Component Structure Analysis

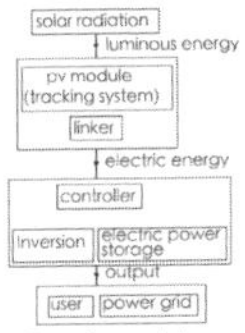

Energy flow analysis

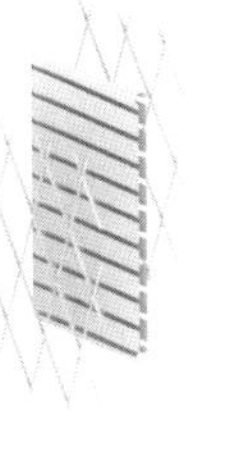

Structural features:
Belongs to three-layer Laminated glass
The structure is outer glass/adhesive film/amorphous silicon battery/adhesive film/inner glass. The outer glass is white glass with good transparency, while the inner glass can be selected in various colors or coated according to needs.

Type:
Full coverage: opaque
Point (line) transparency: Transmitting light from the dots or lines made by the laser.
Louver type: transparent from the battery compartment

Scope of use:
The facade walls and shaded surfaces of the building

Technical parameters:
Conversion efficiency: 5%-8%
Battery thickness: commonly used 3mm
Color: Dark brown
Operating environment temperature: -40-85 ℃
Glass thickness: 7~22mm, various combinations.

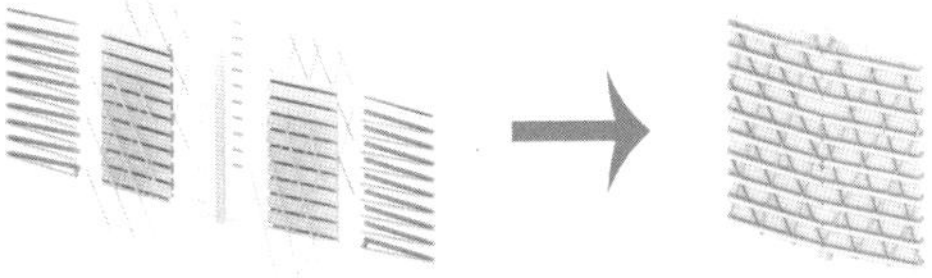

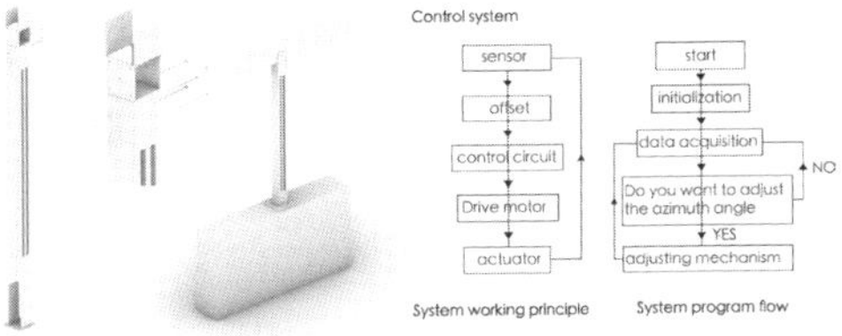

System working principle **System program flow**

Explosion Diagram

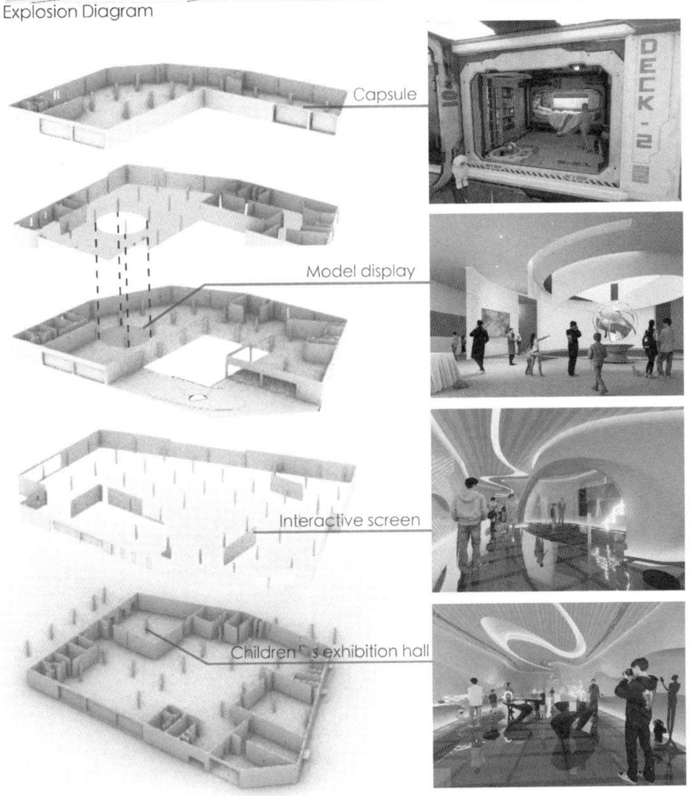

- Capsule
- Model display
- Interactive screen
- Children's exhibition hall

Carbon Emission Indicators

Atrium Louver Glass

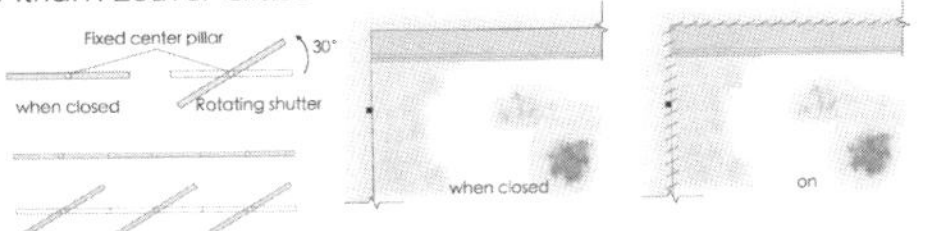

综合奖·入围奖·零碳设计项目
Comprehensive Awards · Nomination · Zero-Carbon Design Project

注册号：101506
Register Number：101506

项目名称：辉晖相映
Entry Title：Symphony of Solar and Snug

作者：蔡资潇、贾英杰、陈子墨、文一博
Authors：Zixiao Cai, Yingjie Jia, Zimo Chen and Yibo Wen

作者单位：东北大学
Authors from：Northeast University

指导教师：刘哲铭
Tutor：Zheming Liu

指导教师单位：东北大学
Tutor from：Northeast University

文化科技馆项目位于广州市增城区科教城内，为更好地服务于群众，提高科普展示的影响力，本方案从人群需求出发，希望打造集非正式课堂、科技沉浸之旅、城市客厅为一体的大众文化科技馆。

结合广州气候特点，为打造舒适低碳环境，本方案以被动设计结合需求空间为起点，通过场地布局和块体关系，打造室外景观和灰空间，降低建筑能耗。为打造沉浸之旅的同时优化大文主建筑环境质量，将沉浸影院架起形成科技馆核心，同时作为绿建技术载体，集促进室内热压通风、优化窗内采光、进行雨水收集和水质净化、绿藻发电展示以及光伏发电和室内灯光调节等多项功能于一体的装置。本方案主要通过光伏发电板进行能源收集，为实现技术与艺术的统一，屋顶采用渐变式单晶硅光伏板，南立面采用光伏变玻璃打造轻盈之感，西立面则着重垂直绿化遮阳与遮风雨的平衡不受置光伏板。此外雨水收益采用叠置净水的方法，提供给市同一个额外的绿色低碳室外展厅。

The cultural science and technology museum project islocated in the science and education city of ZengchengDistrict, Guangzhou, in order to better serve the masses andincrease the influence of science popularization display, thisplan starts from the needs of the crowd and hopes to createa popular culture science and technology museumintegrating informal classrooms, science and technologyimmersion tours, and urban living rooms.

Combined with climate characteristics of Guangzhou,in order to create a comfortable and low-carbonenvironment, this plan takes passive design combined withdemand space as the starting point, and creates outdoorcourtyards and gray spaces through site layout and blockrelationship to reduce building energy consumption. In orderto create an immersive journey and optimize the quality ofthe building environment, the immersive cinema is liftedtoform the core of the science and technology museum, and atthe same time as a green building technology carrier, itintegrates the functions of promoting indoor hot pressureventilation, providing mechanical ventilation, optimizingindoor lighting, conducting rainwater collection and waterpurification, green algae power generation display,photovoltaic power generation and adjusting indoor lightintake. In order to achieve the unity of technology and art,the roof uses graduated monocrystalline silicon photovoltaicpanels, the south facade uses photovoltaic glass to create asense of lightness, and the west facade focuses on thebalance between vertical greening, shading and ventilationwithout photovoltaic panels. In addition, the rain gardenadopts the method of stacking water purification to providethe public with an additional green and low-carbon outdoorexhibition hall.

Design Vision
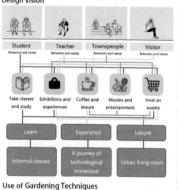

Use of Gardening Techniques

Sun Protection and Insulation for the West Block
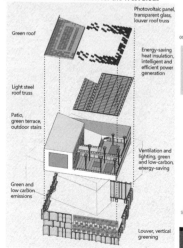

General Plan
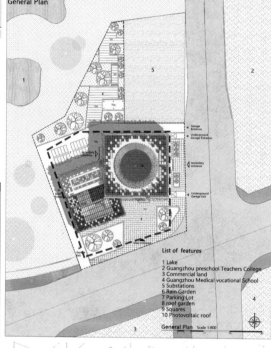

List of features
1 Lake
2 Guangzhou preschool Teachers College
3 Commercial land
4 Guangzhou Medical vocational School
5 Substations
6 Rain Garden
7 Parking Lot
8 roof garden
9 Squares
10 Photovoltaic roof

General Plan Scale 1:800

Block Generation

01 The west side of the site is large area of greenery and low terrain, and the northeast side is a substation
02 Squares are adopted according to the architectural texture of the surrounding plots and rotated according to the trend of the road

03 The height of the building is lowered to echo the landscape while obscuring the substation
04 To further echo the landscape, the masses are retreated to form a roof garden, providing a meeting place.

05 improving the utilization rate of the platform, shading measures were carried out
05 Add shading and photovoltaic measures to the façade and roof

1-1 Profile view 1 : 300

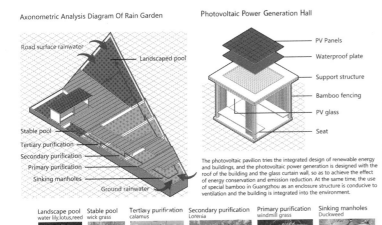

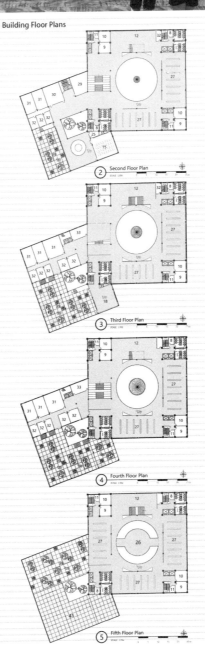
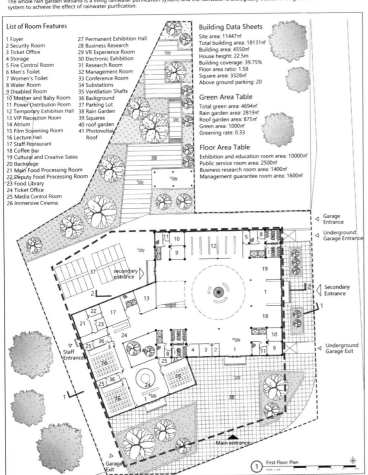

Climatic analysis

The address is located in Zengcheng District, the Pearl River Delta, bordering the the Nanling Mountain Mountains in the north and the South China Sea in the south. The marine climate is characterized by warm and rainy weather, sufficient light and heat, small temperature difference, long summer and short frost period. It is hot in summer and warm in winter.

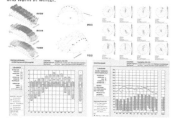

Based on the intuitive chart analysis of meteorological data, we have obtained the climatic conditions that need to be paid attention to in the early stage of design, which makes us pay more attention to temperature, lighting, ventilation and humidity, energy conservation, carbon reduction, and other related aspects of design.

1 Comfort(674 hrs)7.7%
2 Sun Shading of windows(1461 hrs)16.7%
3 Two-Stage Evaporative Cooling(114 hrs)1.3%
4 Internal Heat Gain(1802 hrs)20.6%
5 Passive Solar Direct Gain High mass(474 hrs)5.4%
6 Dehumidification Only(2161 hrs)24.7%
7 Cooling, add Dehumidification if needed(3129 hrs)35.7%
8 Heating, add Humidification if needed(830 hrs)9.5%

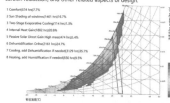

Double Curtain Wall Analysis

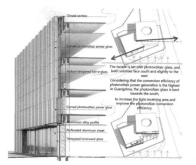

The facade is set with photovoltaic glass, and both volumes face south and slightly to the east.

Considering that the conversion efficiency of photovoltaic power generation is the highest in Guangzhou, the photovoltaic glass is bent towards the south,

to increase the light receiving area and improve the photovoltaic conversion efficiency.

Green Technology Design

Wind Simulation

Summer Outdoor Wind Field

By comparing the three groups of ventilation: wind pressure ventilation, wind pressure and thermal pressure ventilation, wind pressure and thermal pressure ventilation combined with mechanical auxiliary ventilation, and simulating the indoor wind field of the building, the first floor indoor ventilation and cross-sectional ventilation of the control group were obtained.

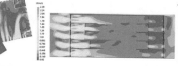
Wind pressure ventilation

Winter Outdoor Wind Field

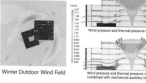
Wind pressure and thermal pressure ventilation

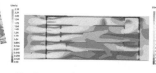
Wind pressure and thermal pressure ventilation

Wind pressure and thermal pressure ventilation combined with mechanical auxiliary ventilation

Wind pressure ventilation: The indoor wind field has a low flow rate

Wind pressure and thermal ventilation: Through the implantation of atrium lighting and ventilation components, the opening of skylights increases the indoor wind speed in both horizontal and cross-sectional areas, achieving thermal ventilation

Wind pressure and heat pressure combined with mechanical auxiliary ventilation: By activating mechanical auxiliary ventilation in the lighting and ventilation components of bathrooms, cinemas and atriums, the flow and ventilation frequency of the wind field are significantly increased, and the indoor ventilation effect is good.

Wind pressure and thermal pressure ventilation combined with mechanical auxiliary ventilation

Indoor at Summer and Winter Solstice

By simulating indoor temperature conditions in summer and winter the indoor temperature conditions of the first floor, standard floor, and top floor of the building were obtained. Due to the presence of atrium lighting and ventilation components, the opening of the skylight caused different changes in indoor temperature around the atrium lighting and ventilation components, and the temperature of each floor also varied. The chart shows that indoor temperature increases from bottom to top in summer and decreases from bottom to top in winter.

Indoor Thermal Comfort Level

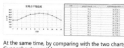

At the same time, by comparing with the two charts of Guangzhou's monthly average outdoor temperature and indoor thermal comfort temperature range, it was found that the temperatures of each layer were within the comfort range. The proportion of time required for buildings to meet the thermal comfort zone is 67.44%, which is based on the requirements of Article 5.2.9 (1) of the "Green Building Evaluation Standard" GB/T50378-2019 regarding indoor thermal and humid environments. The area weighted average is calculated based on the calculated values of each main functional room in the building.

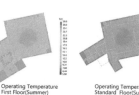

Operating Temperature First Floor(Summer)
Operating Temperature Standard Floor(Summer)
Operating Temperature Top Floor(Summer)

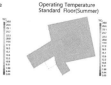
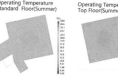

Operating Temperature First Floor(Winter)
Operating Temperature Standard Floor(Winter)
Operating Temperature Top Floor(Winter)

South elevation view 1: 400

Indoor Illuminance Analysis

Indoor Illuminance (Without Skylight) First Floor

Indoor Illuminance (With Skylight) First Floor

Indoor Illuminance (Without Skylight) Standard Floor

Indoor Illuminance (With Skylight) Standard Floor

Indoor Illuminance (Without Skylight) Top Floor

Indoor Illuminance (With Skylight) Top Floor

Simulate indoor lighting in summer, compare and analyze whether there are atrium lighting and ventilation components, and divide them into skylight group and non skylight group. The figure shows the lighting conditions and comparative analysis of the first floor, standard floor, and top floor of the building.

The skylight group has a certain area of transparent glass on the top layer, allowing light to enter the corespace of the building through the skylight, creating a softer and brighter environment, and significantly improving the lighting situation.

Atrium Lighting and Ventilation Components

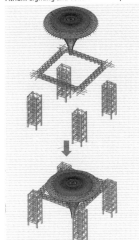
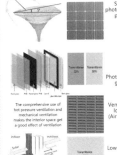
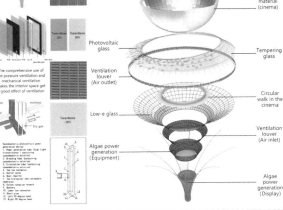
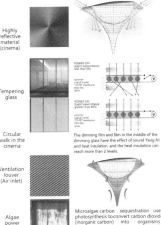

Microalgae carbon sequestration uses photosynthesis to convert carbon dioxide (inorganic carbon) into organisms (organic carbon and oxygen), which will be the core exhibition technology of the Science and Technology Museum.

2-2 Profile Perspective

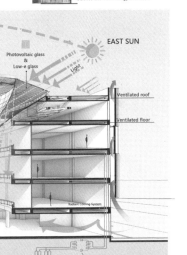

综合奖・入围奖・零碳设计项目
Comprehensive Awards · Nomination · Zero-Carbon Design Project

注册号：101507
Register Number：101507

项目名称：风之谷
Entry Title：The Valley of Wind

作者：岳琮达、胡浪浪、刘嘉琪、申佳梅
Authors：Congda Yue, Langlang Hu, Jiaqi Liu and Jiamei Shen

作者单位：沈阳建筑大学
Authors from：Shenyang Jianzhu University

指导教师：高畅、侯静
Tutors：Chang Gao and Jing Hou

指导教师单位：沈阳建筑大学
Tutors from：Shenyang Jianzhu University

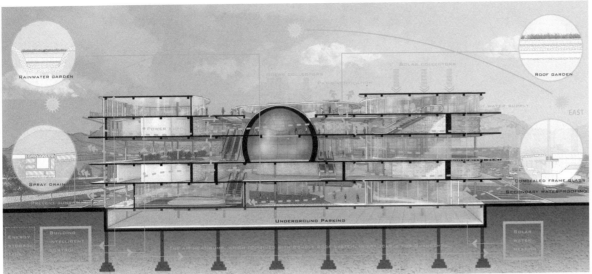
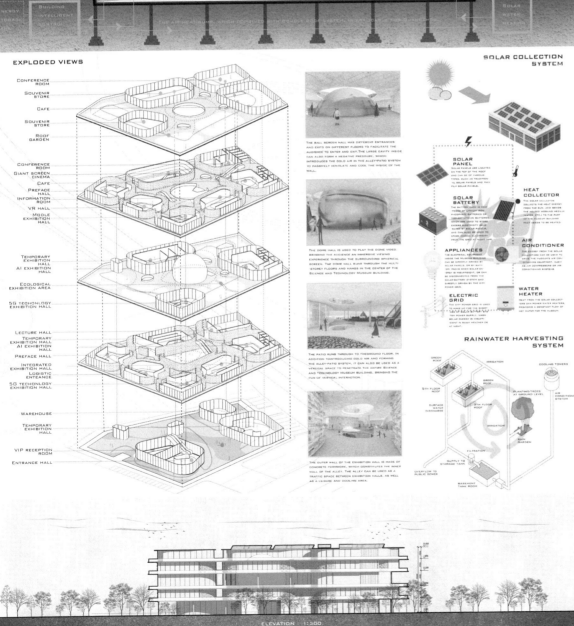

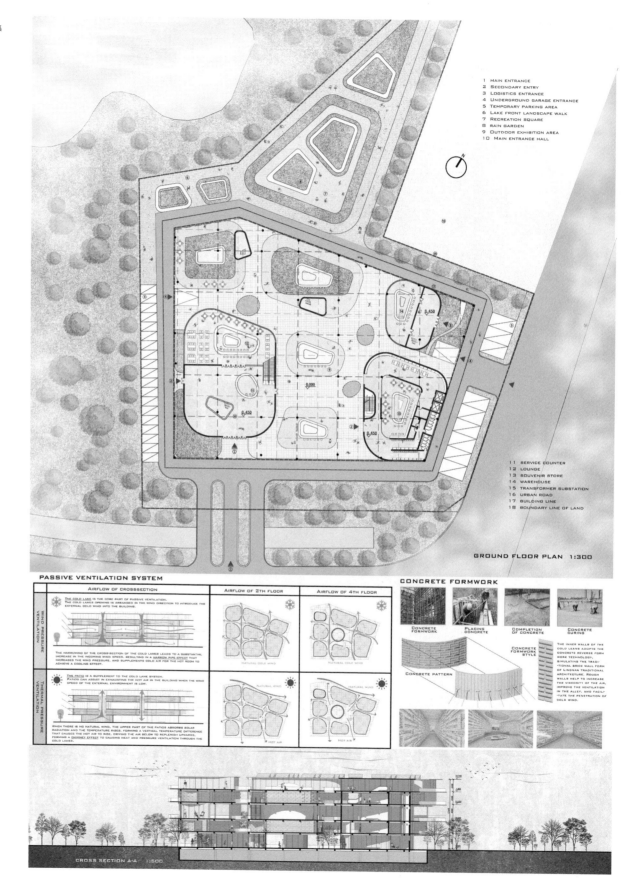

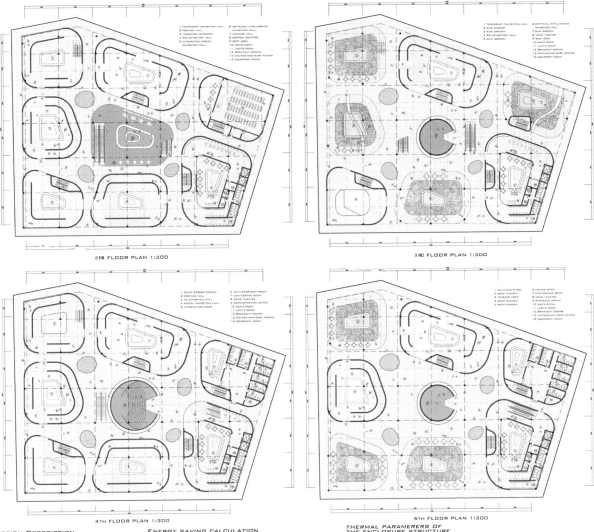

DESIGN DESCRIPTION

IN TERMS OF FUNCTIONALITY, A METHOD OF DISASSEMBLING AND REARRANGING FUNCTIONS IS ADOPTED TO SORT OUT THE FUNCTIONS OF THE BUILDING.

IN TERMS OF SHAPE CONTROL, A GRID ANALYSIS METHOD WAS ADOPTED, WITH A MODULUS OF 6*6 AS THE BASIC CONTROL LINE FOR THE DESIGN.

IN TERMS OF VENTILATION EFFECT, THIS BUILDING ADOPTS THE UNIQUE VENTILATION SYSTEM OF COLD ALLEYS, AND COURTYARDS IN LINGNAN RESIDENTIAL AREAS. THIS MAKES THE BUILDING FORM A STEREOSCOPIC CIRCULATING VENTILATION SYSTEM IN BOTH HORIZONTAL AND VERTICAL DIRECTIONS.

THE BUILDING USES SOLAR COLLECTORS AND PHOTOVOLTAIC CURTAIN WALLS ON BOTH THE ROOF AND HE GLASS ON THE EXTERIOR SURFACE OF THE BUILDING, AND IS SUPPLIED TO THE BUILDING THROUGH A CENTRAL POWER CONTROL SYSTEM.

IN DEALING WITH SUNLIGHT, THE BUILDING ADOPTS A THREE METERS OVERHANG TO BLOCK OUT SUNLIGHT TO COOL THE INTERIOR.

AT THE SAME TIME, TO AVOID EXPOSURE TO THE WEST, WE HAVE ADDED WINDOW-SHADES INSIDE THE BUILDING.

ENERGY SAVING CALCULATION

	unit	numerical value
Effective area of photovoltaic glass	m²	1,944
Solar collector plate area	m²	2,478
Total solar area	m²	4,422
Annual solar energy production	kilowatt hour	807,015

CALCULATION OF BUILDING ENERGY CONSUMPTION

	unit	numerical value
Energy consumption of ordinary building	kilowatt hour	906,360
Energy consumption of this building	kilowatt hour	99,345

EACH ENERGY CONSUMPTION ITEM

	unit	numerical value
Heating set	kilowatt hour	95,459
Air-conditioner set	kilowatt hour	387,517
Lighting	kilowatt hour	426,373
All-year energy consumption	kilowatt hour	906,360

THERMAL PARAMETERS OF THE ENCLOSURE STRUCTURE

		unit	designed building	reference building
Roof		W/(m²·K)	0.71	0.8
Exterior wall		W/(m²·K)	1.52	1.5
Exterior window	K east	W/(m²·K)	2.3	2.9
	south	W/(m²·K)	2.4	2.6
	west	W/(m²·K)	2.5	2.7
	north	W/(m²·K)	2.4	3.1
	SHGC east		0.32%	0.37%
	south		0.35%	0.37%
	west		0.32%	0.37%
	north		0.32%	0.47%
Roof light	K	W/(m²·K)	2.4	3.1
	SHGC	SHGC	0.29	0.49
	area proportion		0.03	0.19

BUILDING INDICATORS

	unit	numerical value
Floor space	m²	18567.31
Building land area	m²	5575.75
Floor area ratio	m²	3.33
Greening rate	/	27.66%

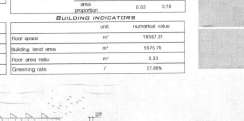

CELL PRESSURE CFD ANALYSIS -80–250PA

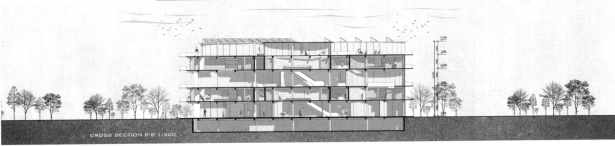

CROSS SECTION 6-6 1:300

综合奖・入围奖・零碳设计项目
Comprehensive Awards・Nomination・Zero-Carbon Design Project

注册号：101534
Register Number：101534

项目名称：风巷・树影
Entry Title：Wind Alley, Trees Shadow

作者：王秉仁、李明远、常昊、梁浩楠、张忠瑞、徐端、刘轩彤
Authors：Bingren Wang, Mingyuan Li, Hao Chang, Haonan Liang, Zhongrui Zhang, Duan Xu and Xuantong Liu

作者单位：天津大学
Authors from：Tianjin University

指导教师：郭娟利、李伟
Tutors：Juanli Guo and Wei Li

指导教师单位：天津大学
Tutors from：Tianjin University

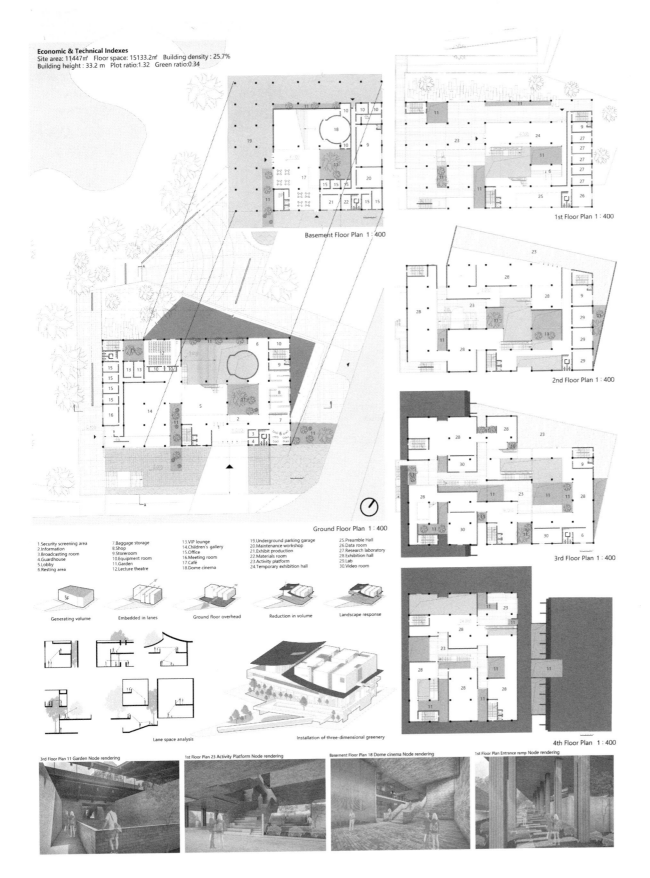

Wall pattern

Analysis of photovoltaics and carbon emissions

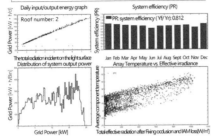

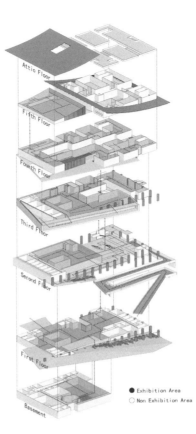

综合奖·入围奖·零碳设计项目
Comprehensive Awards · Nomination · Zero-Carbon Design Project

注册号：101539
Register Number：101539

项目名称：风正一帆悬
Entry Title：The Sail Hangs Straight Amid the Wind

作者：王馨可、张月阳
Authors：Xinke Wang and Yueyang Zhang

作者单位：中国矿业大学
Authors from：China University of Mining and Technology

指导教师：马全明、段忠诚、邵泽彪
Tutors：Quanming Ma, Zhongcheng Duan and Zebiao Shao

指导教师单位：中国矿业大学
Tutors from：China University of Mining and Technology

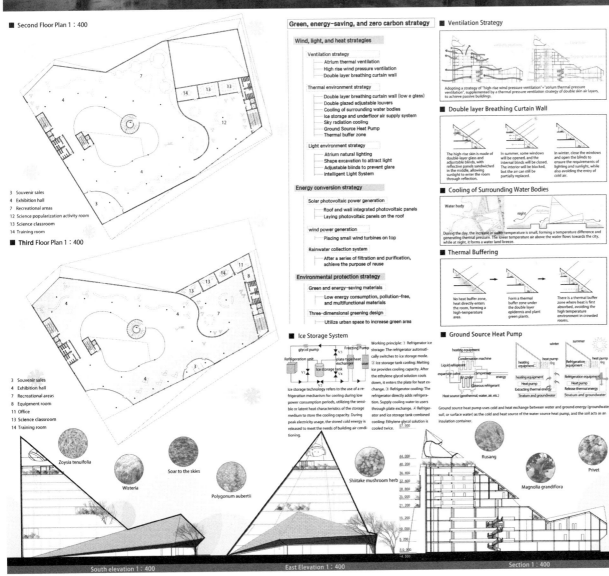

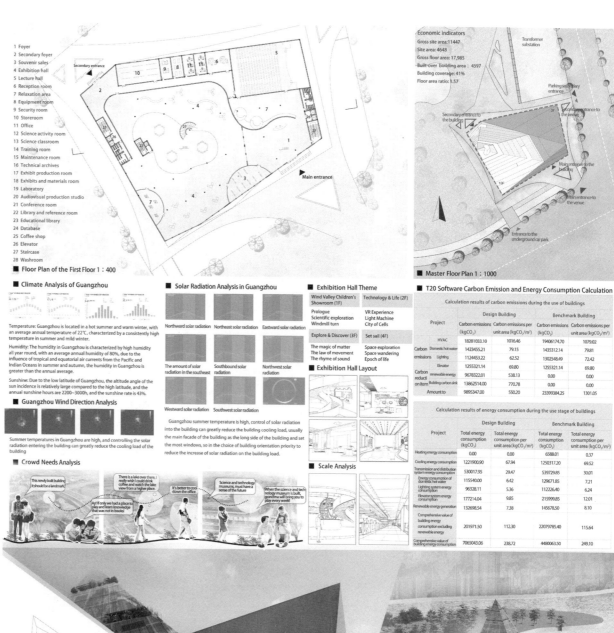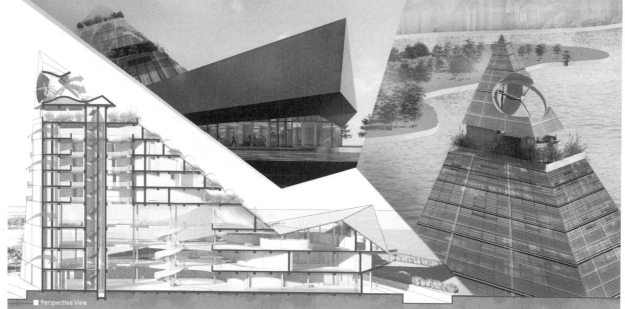

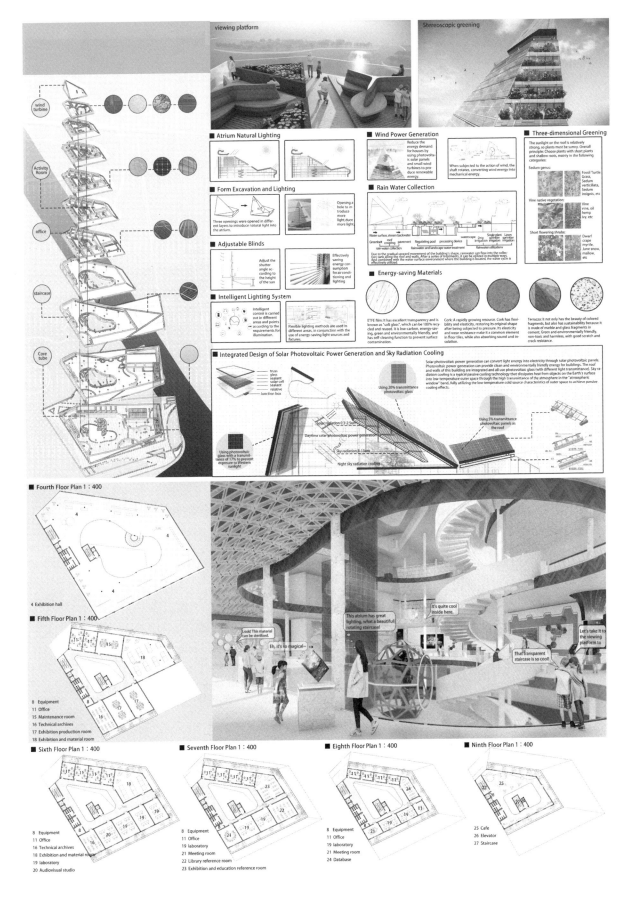

综合奖·入围奖·零碳设计项目
Comprehensive Awards · Nomination · Zero-Carbon Design Project

注册号：101552
Register Number：101552

项目名称：创梦盒伙人
Entry Title：Dream Partners

作者：丁佳钰、吴欣然、杨栋、易凡、梁红欣
Authors：Jiayu Ding, Xinran Wu, Dong Yang, Fan Yi and Hongxin Liang

作者单位：南京工业大学
Authors from：Nanjing Tech University

指导教师：罗靖、陈建标、刘静萍
Tutors：Jing Luo, Jianbiao Chen and Jingping Liu

指导教师单位：南京工业大学
Tutors from：Nanjing Tech University

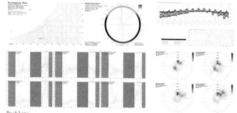

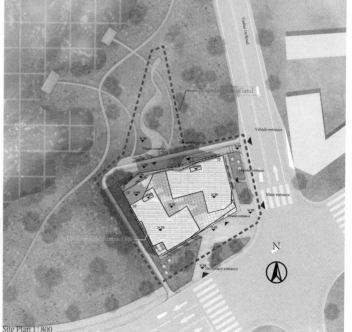

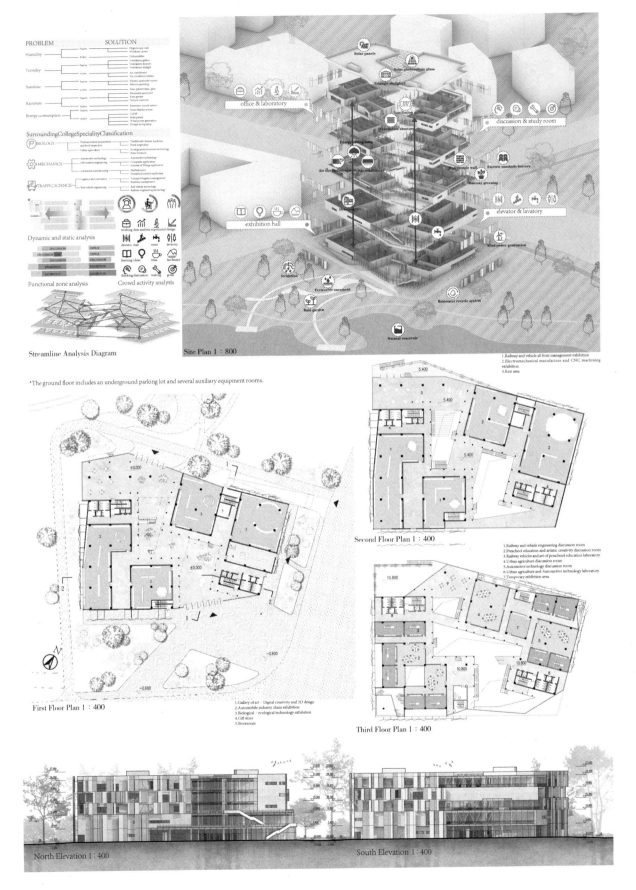

Architectural Skin Perspective

Atrium Perspective

Stepping Terrace Perspective

Fourth Floor Plan 1∶400

Fifth Floor Plan 1∶400

Passive Design Strategies

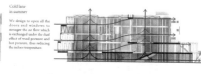

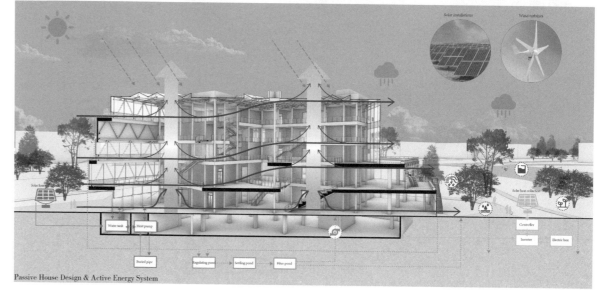
Passive House Design & Active Energy System

Window Analysis Diagram

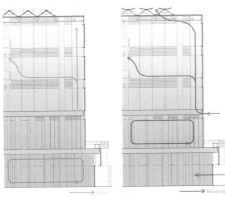

Windows closed · Windows open

Skylight Construction

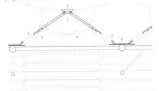

Wall construction diagram

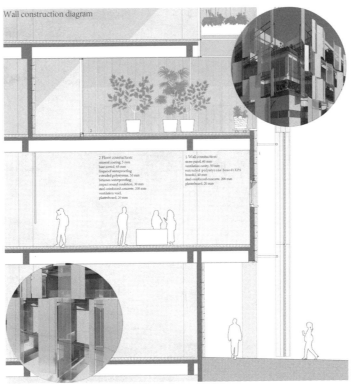

1-1 Section 1 : 400

2-2 Section 1 : 400

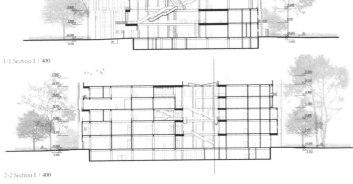

Green Construction Evaluation Form

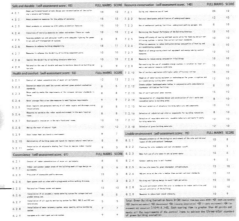

Night view of the lake

综合奖·入围奖·零碳设计项目
Comprehensive Awards · Nomination · Zero-Carbon Design Project

注册号：101617
Register Number：101617

项目名称：无界方舟
Entry Title：The Unbounded Ark

作者：周珺童、宁可、何毅贤、梅颢耀、郭映雪、杨佳成、郝建科、陈佳言
Authors：Juntong Zhou, Ke Ning, Yixian He, Haoyao Mei, Yingxue Guo, Jiacheng Yang, Jianke Hao and Jiayan Chen

作者单位：西安交通大学、华南农业大学、香港中文大学
Authors from：Xi'an Jiaotong University, South China Agricultural University and The Chinese University of Hong Kong

指导教师：王海旭、徐怡珊、虞志淳
Tutors：Haixu Wang, Yishan Xu and Zhichun Yu

指导教师单位：西安交通大学
Tutors from：Xi'an Jiaotong University

Climatic SImulation

"Climate Space" Analyze

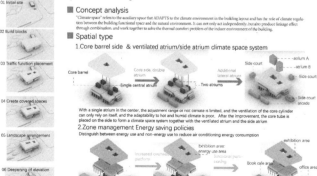

Climatic SImulation

Passive strategy:
1. The optimal orientation of the building is south;
2. In the passive strategy, natural ventilation is strengthened, especially the southeast wind in summer;
3. To prevent overheating in summer, it is necessary to pay attention to shading strategies

Active strategies:
1. solar radiation is mainly rich, solar photovoltaic power generation can be used;
2. Rainwater collection can be organized,and sponge city design can be combined with wetland landscape

Section

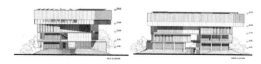

Exploded Drawing

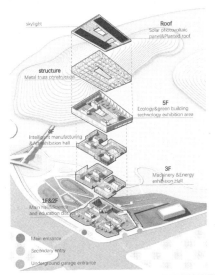

Floor Plan

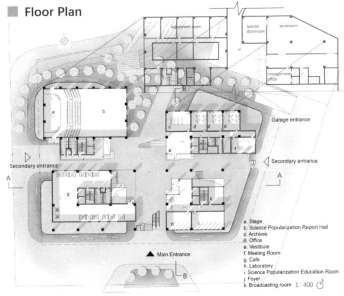

a. Stage
b. Science Popularization Report Hall
c. Archives
d. Office
e. Vestibule
f. Meeting Room
g. Cafe
h. Laboratory
i. Science Popularization Education Room
j. Foyer
k. Broadcasting room 1 : 400

Rainwater Cycle

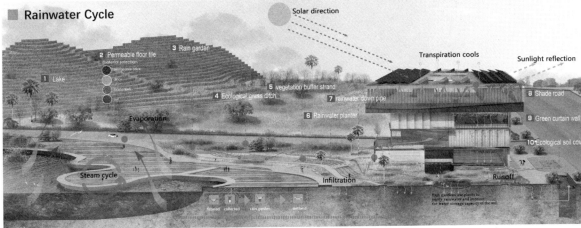

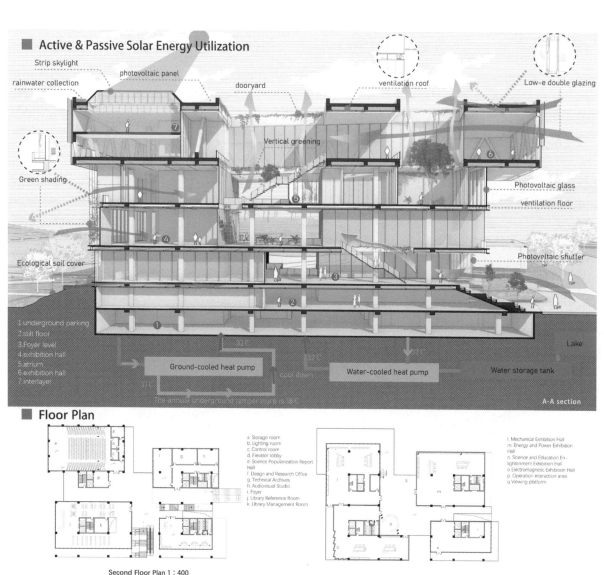

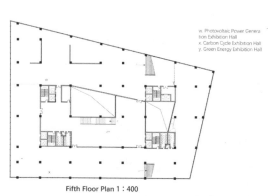

Passive Solar Technology

Carbon Emission & Energy Consumption Calculation

Building carbon emission index table
The initial parameters for energy consumption calculation of various functional rooms are as shown in the table below.

Room Functions	Temperature (°C)	Relative Humidity (%)	Temperature (°C)	Relative Humidity (%)	Minimum Fresh Air Volume	Occupancy Density (people/m²)	Lighting Power (W/m²)	Equipment Power (W/m²)
Lobby Foyer	26	60	20	40	20	0.02	15	0
Office	26	60	20	40	30	0.1	18	13
Restroom or Bathroom	26	70	18	40	20	0.1	6	0
Multipurpose Hall	26	65	20	40	20	0.25	11	0
Exhibition Hall or Gallery	27	60	18	40	20	0.71	11	20
Store or Shop	27	65	20	40	20	1	12	13

The energy consumption calculation results for various functional rooms are as shown in the table below.

Room Functions	Fresh Air (m³)	Lighting (Watts)	Equipment (Watts)	Number of People
Lobby Foyer	1868	7005	0	9.34
Office	374638	23094	16679	1248.788667
Restroom or Bathroom	584	1752	0	29.2
Multipurpose Hall	5260	11160	0	2625
Exhibition Hall or Gallery	90034	68970	125400	4451.7
Store or Shop	7900	4260	4145	395
Total	1206914	115751	146624	8165.826667

The building operational energy consumption, photovoltaic generation, and carbon sequestration by plants are shown in the table below.

HVAC Energy Consumption (kW·h)	Lighting and Equipment Energy Consumption (kW·h)	Total Energy Consumption (kW·h)	Photovoltaic Generation (kW·h)	Total Carbon Emissions (kg)	Photovoltaic Emission Reduction (kg)	30% Greening Carbon Reduction (kg)
1017957.7	2776175	3794192.7	1082673	860010.383	1079404.98	287973.115

Simulation analysis specification

The HVAC energy consumption section includes the annual cooling and heating loads as well as latent heat loads of the building. The building envelope is covered with insulation materials, and the glass is triple-layer vacuum low-e glass. The air conditioning system is adjusted on an hourly basis, and a high-performance ground source heat pump is used for the heat pump unit to reduce energy consumption. Lighting and equipment loads are calculated based on the standard values provided by regulations. The total carbon emissions are calculated by multiplying the total energy consumption by the carbon emission factor of the southern regional grid. The photovoltaic emission reduction is calculated by multiplying the photovoltaic generation by the carbon reduction factor. The source of photovoltaic power generation comes from the rooftop photovoltaic panels and photovoltaic shading panels on the rooftop side. After 30% greening, carbon emissions are calculated according to the standards for zero-carbon buildings. For small public buildings, 30% greening is allowed as a carbon sink. In other words, if the current photovoltaic emission reduction exceeds the carbon emissions after greening, it can achieve zero-carbon operation.

Open-Loop External Circulation System Breathable Curtain Wall

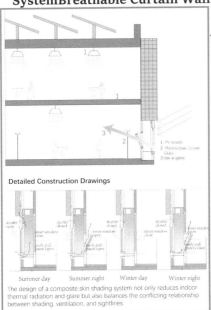

The design of a composite skin shading system not only reduces indoor thermal radiation and glare but also balances the conflicting relationship between shading, ventilation, and sightlines.

Open-Loop External Circulation System Breathable Curtain Wall

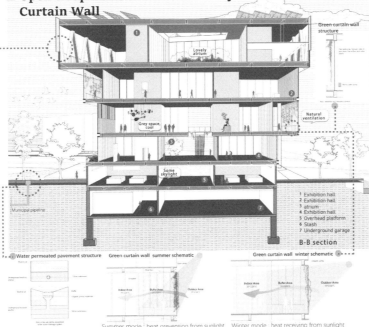

Illuminance Analysis

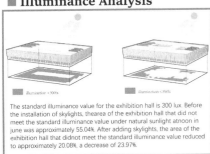

The standard illuminance value for the exhibition hall is 300 lux. Before the installation of skylights, the area of the exhibition hall that did not meet the standard illuminance value under natural sunlight at noon in june was approximately 55.04%. After adding skylights, the area of the exhibition hall that didnot meet the standard illuminance value reduced to approximately 20.08%, a decrease of 23.97%.

Intelligent Building Control

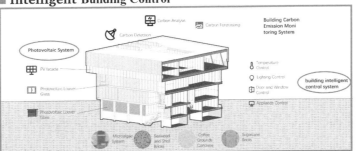

综合奖·入围奖·零碳设计项目
Comprehensive Awards · Nomination · Zero-Carbon Design Project

注册号：101638
Register Number：101638

项目名称：风科谷
Entry Title：Technical Air Flue

作者：宋坤阳、刘奕彤、信彦涛、李子涵
Authors：Kunyang Song, Yitong Liu, Yantao Xin and Zihan Li

作者单位：沈阳建筑大学
Authors from：Shenyang Jianzhu University

指导教师：张龙巍、鞠叶欣
Tutors：Longwei Zhang and Yexin Ju

指导教师单位：沈阳建筑大学
Tutors from：Shenyang Jianzhu University

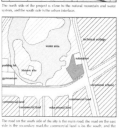

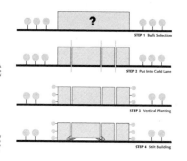

■ PRE-STUDY

According to the software simulation, statistical analysis, we found that the width of the cold aisle and the size of the wind speed show a parabola, and the wind speed is the largest between 2M-2.5M, at about 5m/s. The optimal wind speed for human body is generally a breeze of three levels, so a ventilation item of about 2M can be used with a breeze gallery of 5M or more.

In terms of height, the wind speed curves almost overlap with no change at 16M and 20M, and show a larger change at 24M.

We use the "cold lane" principle, which is combined with the cavity, to obtain the optimal width of the gap for maximum air velocity through data simulation.
Software: PHOENICS

Guangdong summer wind was used for simulation analysis, and the value was set as S-S-E with a wind speed of 2.3m/s.
We measured 1M, 2M, 2.5M, 3M, 4M, and 5M, six pitch widths; and three heights in each pitch, 16M, 20M, and 24M, respectively.

We conducted another angular simulation, taking a gap of 3M width for measurement, and tested the size of three kinds of openings, and finally found that the wind speed in the gap is the largest when the two body blocks are parallel, and the opening will slow down the wind speed in the second half.

CONCLUSION

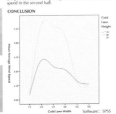

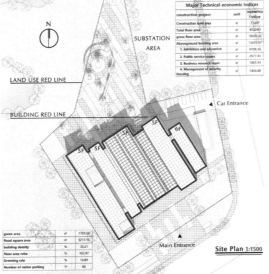

Site Plan 1:1500

■ GROUND FLOOR OVERHEAD

How to partially elevate the ground floor?
When the ground floor is partially elevated, wind can flow better into the building.

Night
The large space on the ground floor closes to form a closed space.

Daytime
The façade unit opens to create a bottom overhead to allow wind passage.

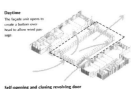

Self-opening and closing revolving door

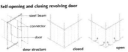

■ STREAMLINE ANALYSIS

Clear flow lines are necessary for the science and technology museum, and the project is divided into three flow lines.

■ TECHNICAL ANALYSIS

The project utilizes cold alley ventilation to achieve smooth ventilation throughout the entire building and bring about cooling.

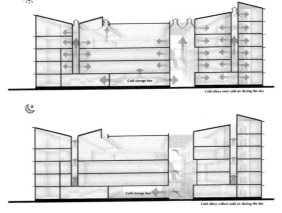

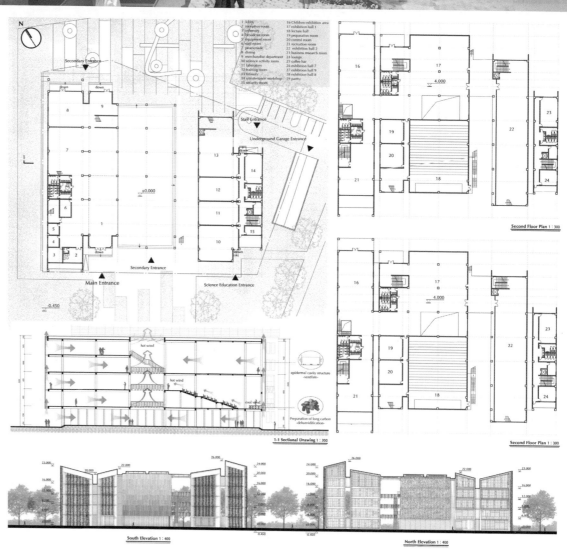

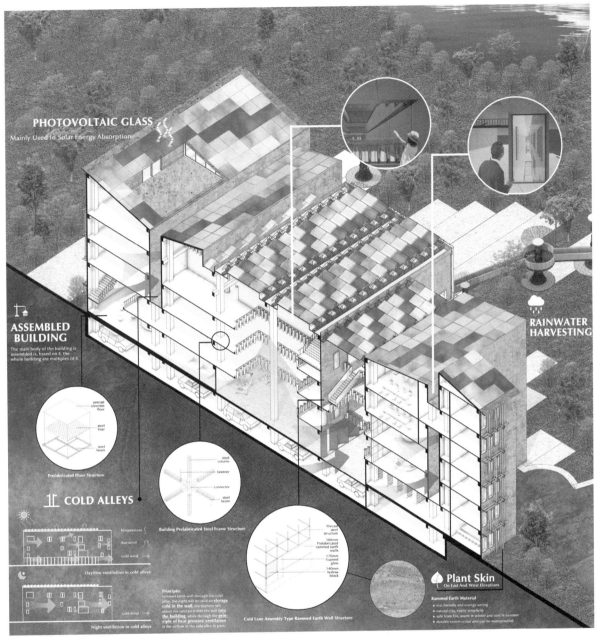

综合奖・入围奖・零碳设计项目
Comprehensive Awards・Nomination・Zero-Carbon Design Project

注册号：101690
Register Number：101690
项目名称：山谷生长
Entry Title：Growing Valleys
作者：吕萌、张万瑜、胡诗媛
Authors：Meng Lyu, Wangyu Zhang and Shiyuan Hu
作者单位：北京交通大学
Authors from：Beijing Jiaotong University
指导教师：胡映东、杜晓辉
Tutors：Yingdong Hu and Xiaohui Du
指导教师单位：北京交通大学
Tutors from：Beijing Jiaotong University

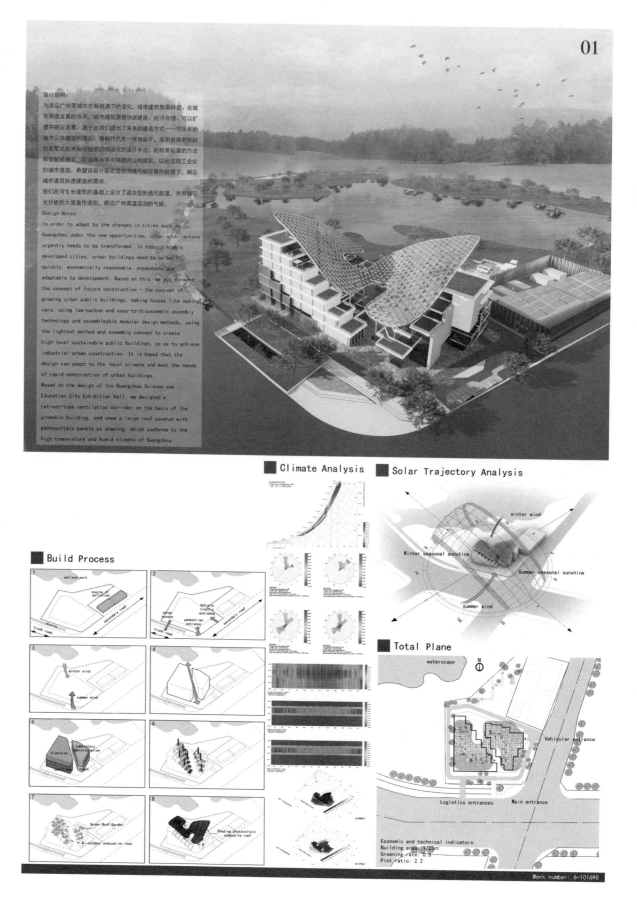

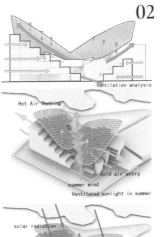

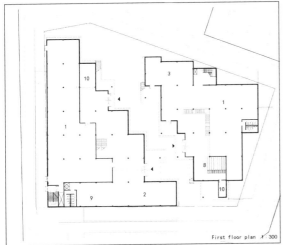
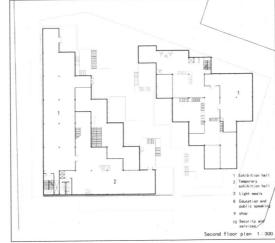
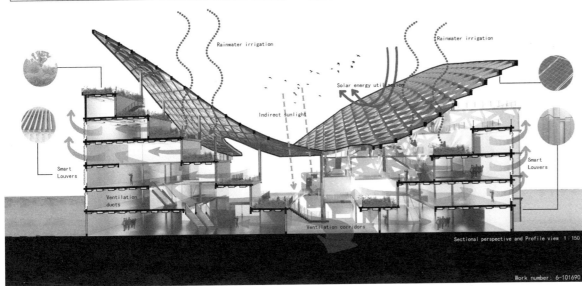

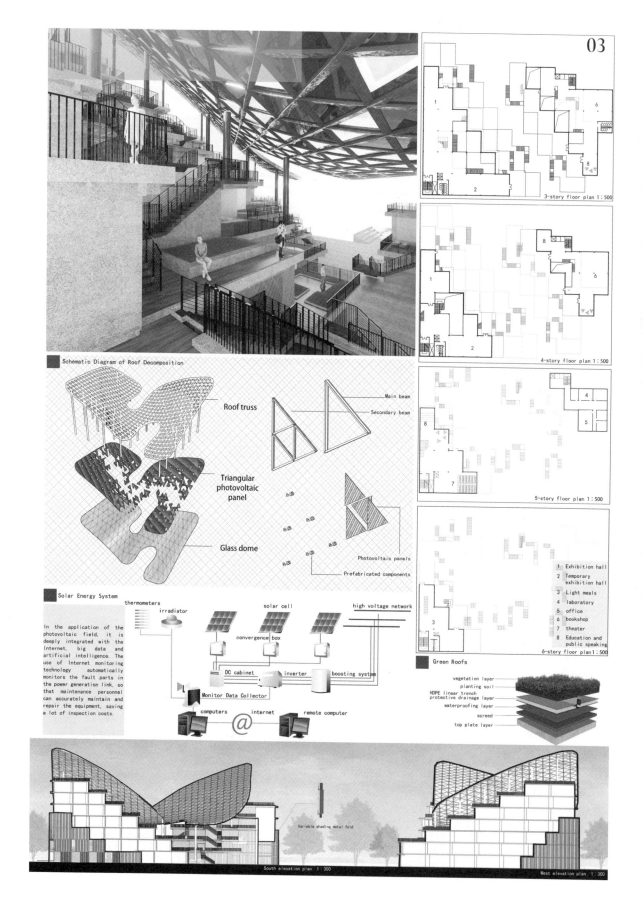

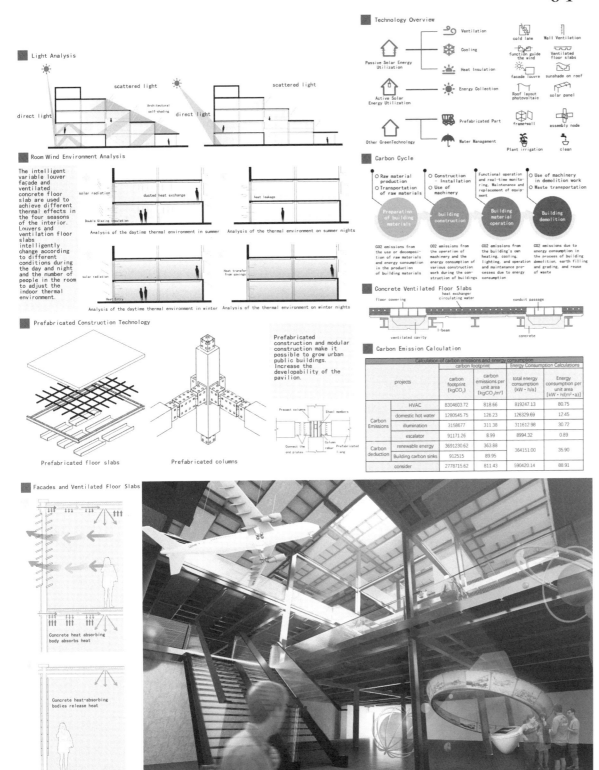

综合奖・入围奖・零碳设计项目
Comprehensive Awards · Nomination · Zero-Carbon Design Project

注册号：101716
Register Number：101716

项目名称：梯山栈谷
Entry Title：Terraced Mountain Plank and Valley

作者：马千里、李睿婧、郑翔书、刘鸿霖、刘玉灏
Authors：Qianli Ma, Ruijing Li, Xiangshu Zheng, Honglin Liu and Yuhao Liu

作者单位：华南理工大学
Authors from：South China University of Technology

指导教师：赵立华
Tutor：Lihua Zhao

指导教师单位：华南理工大学
Tutor from：South China University of Technology

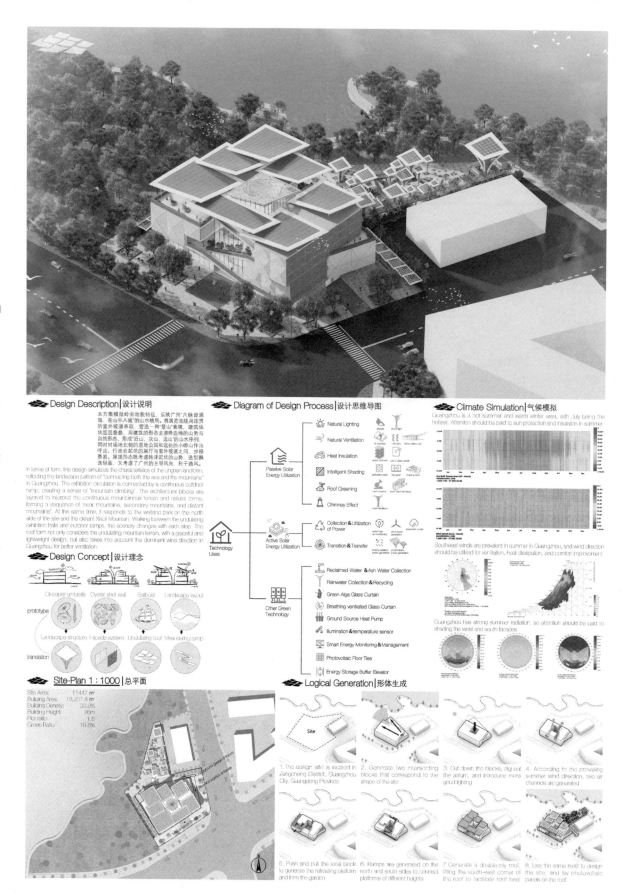

High tech umbrella | 科技之伞

Rainwater collection & evaporation cooling | 雨水收集和蒸发降温

Ecological Park | 生态公园

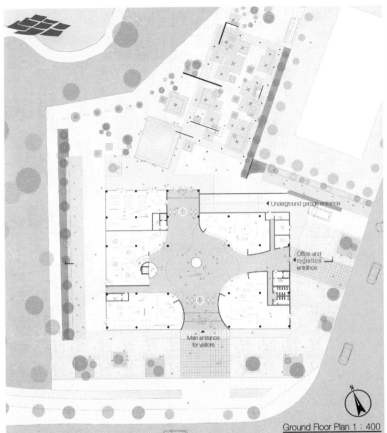
Ground Floor Plan 1:400

1 Ticket and Information
2 Eco-Showroom: Time Machine, Low Carbon Future
3 Children's Gallery: Star Exploration, Space Odyssey
4 City Showroom: Technology Engine, Wisdom Zangcheng
5 Sustainable Showroom: Green Box, Protect the Planet
6 Green Building Technology Center Showroom
7 Atrium Sightseeing Elevator
8 Souvenir Shop
9 Fire Control Room
10 Infirmary
11 Art Café & Lunch
12 Waiting Area
13 Resting Area
14 Youth Reading Room
15 Low Carbon Gallery I: New Materials
16 Low Carbon Gallery II: Renewable Energy
17 Low Carbon Gallery III: New Energy Vehicles
18 Function Room
19 Future Showroom: Welcome Back to Nature
20 Auditorium
21 Dome Theater
22 Data Room
23 Exhibition Resource Room
24 Research Labs
25 Terrace Garden
26 Meeting Rooms
27 Offices

Second Floor Plan 1:500   Third Floor Plan 1:500

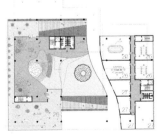

Fourth Floor Plan 1:500   Five floor plan 1:500

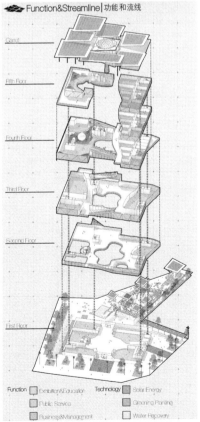
Function & Streamline | 功能和流线

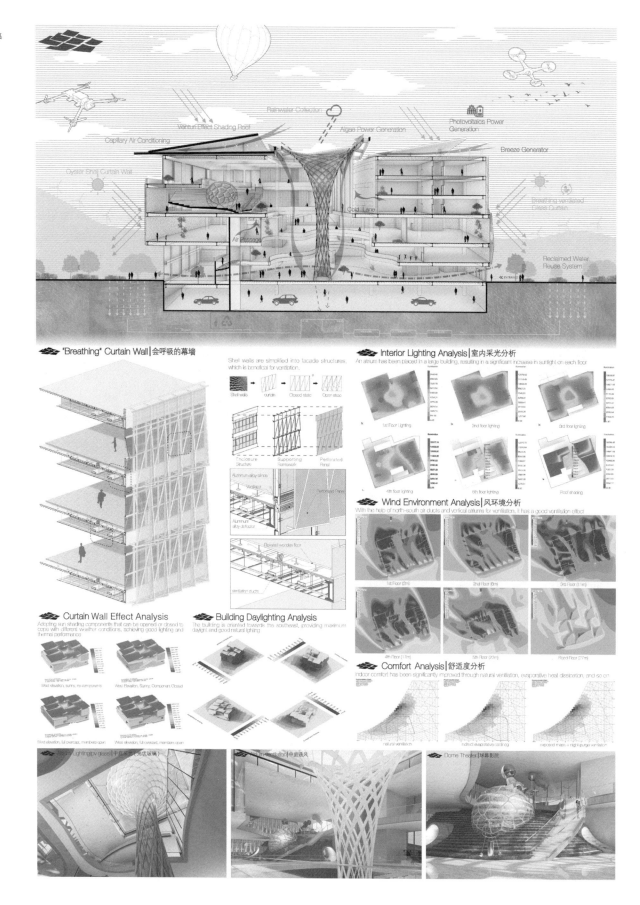

Active Solar Technology | 主动式太阳能技术

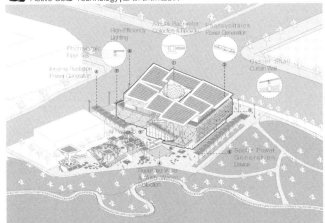

High Tech Umbrella | 科技之伞

Plant Footprint & South Facade | 植物分析 & 南立面

Carbon Emission Calculation Report | 碳排放计算报告

	Calculation Result of Carbon Emission in Building Use Stage			
Item	Designed Building		Reference Building	
	Carbon Emission (kgCO₂)	Carbon Emissions per-unit Area (kgCO₂/m²)	Carbon Emission (kgCO₂)	Carbon Emissions per unit Area (kgCO₂/m²)
Carbon Emission Term	Heating Ventilation Air Conditioning	9519415.20	521.40	22972485.52	1258.26
	Domestic Hot Water	675281.35	36.99	2053793.33	112.49
	Illumination	5499702.38	301.23	7714264.00	422.53
	Elevator	36583.32	2.00	133342.16	7.30
Carbon Reduction Term	Renewable Energy	6524862.82	357.38	0.0	0.0
	Building Carbon Sink	3778904.75	205.86	0.0	0.0
Total		5427214.68	298.39	32873885.00	1800.58

Review of Carbon Emission Intensity Index in Building Use Stage			
Item	Numerical Value	Standard Requirement	Meet the Requirements or Not
Carbon Emission Intensity Reduction (kgCO₂/m²·a)	33.17	≥7	Yes
Carbon Intensity Reduction Rate (%)	84.43	≥40	Yes

Calculation Results of Energy Production from Renewable Energy Sources in the Design Building use stage							
Photovoltaic Power Generation kW·h/(m²·a)	Wind Power Generation kW·h/(m²·a)	Solar Heating/ Cooling kW·h/(m²·a)	Solar Domestic Hot Water kW·h/(m²·a)	Ground Source Heat Pump kW·h/(m²·a)	Air Source Heat Pump kW·h/(m²·a)	Air Source Heat Pump Domestic Hot Water kW·h/(m²·a)	Biomass kW·h/(m²·a)
35.25	0.01	16.26	1.4	0.0	0.0	0.0	0.0

Tridimensional Virescence | 立体绿化

综合奖·入围奖·零碳设计项目
Comprehensive Awards · Nomination · Zero-Carbon Design Project

注册号：101721
Register Number：101721

项目名称：光隐南园
Entry Title：Lingnan Garden Behind the Solar PV Panel

作者：王芊瑶、张皓麒、周小琴、龙思彤、李文苑
Authors：Qianyao Wang, Haoqi Zhang, Xiaoqin Zhou, Sitong Long and Wenyuan Li

作者单位：广州大学
Authors from：Guangzhou University

指导教师：庞玥、李丽
Tutors：Yue Pang and Li Li

指导教师单位：广州大学
Tutors from：Guangzhou University

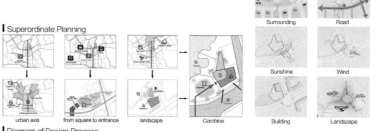

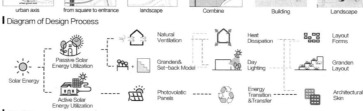

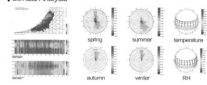

Local Effect Diagram

Overhead gardens · Roof garden · Outdoor landscaped gardens

Block Generation

- Building area & Tranformer station
- Building & Garden
- Wind flows through the mid courtyard
- Dig the box for more balcony
- Cut back shape for flowing and beauty
- The infiltrate and blossom of garden

Ground Floor Plan

1. Visitor Entrance Lobby
2. Relaxation Tea Room
3. Temporary Exhibition Hall
4. Multifunctional Hall
5. Transportation Hall
6. Staff Lobby
7. Management Office
8. Storeroom
9. Exit of the Exhibition Area
10. Underground Garage Entrance

Ground Floor Plan 1 : 350

Lingnan Gardens Analysis

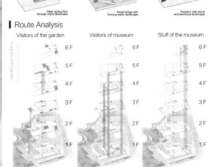

Route Analysis

Visitors of the garden · Visitors of museum · Stuff of the museum

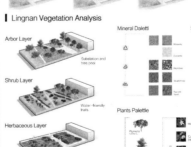

Lingnan Vegetation Analysis

- Arbor Layer — Substation and tree pool
- Shrub Layer — Water-friendly trails
- Herbaceous Layer — Roof garden
- Aquatic plant Layer — Bridge

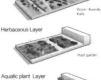

Mineral Dalettl · Size

Plants Palettle

Vegetation Distribution

Stormwater runoff

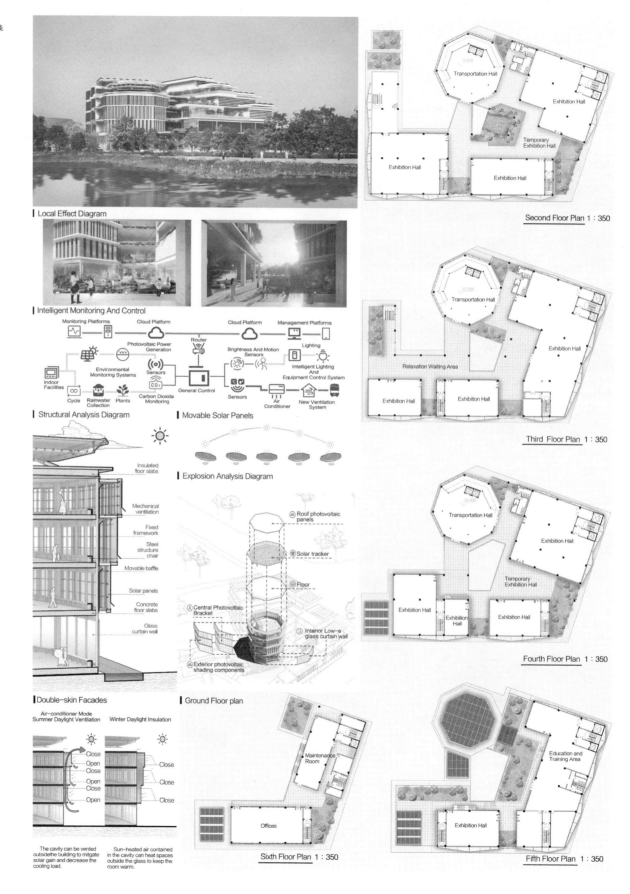

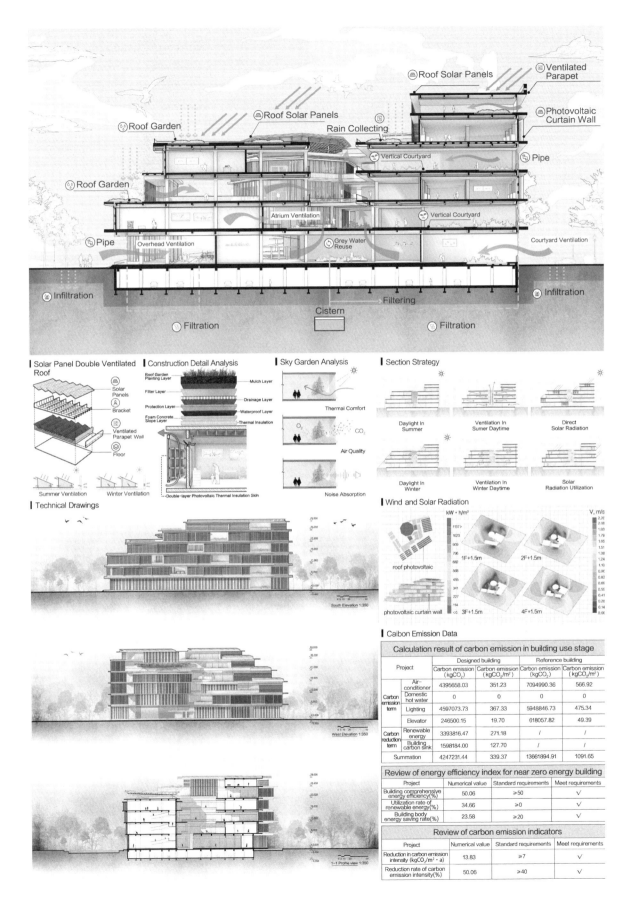

综合奖・入围奖・零碳设计项目
Comprehensive Awards · Nomination · Zero-Carbon Design Project

注册号：101879
Register Number：101879

项目名称：旋・浮
Entry Title：Spiral, Suspension

作者：吴浩朴、赵子慧、叶萌、王钰鑫
Authors：Haopu Wu, Zihui Zhao, Meng Ye and Yuxin Wang

作者单位：河南大学
Authors from：Henan University

指导教师：康永基、宗慧宁
Tutors：Yongji Kang and Huining Zong

指导教师单位：河南大学
Tutors from：Henan University

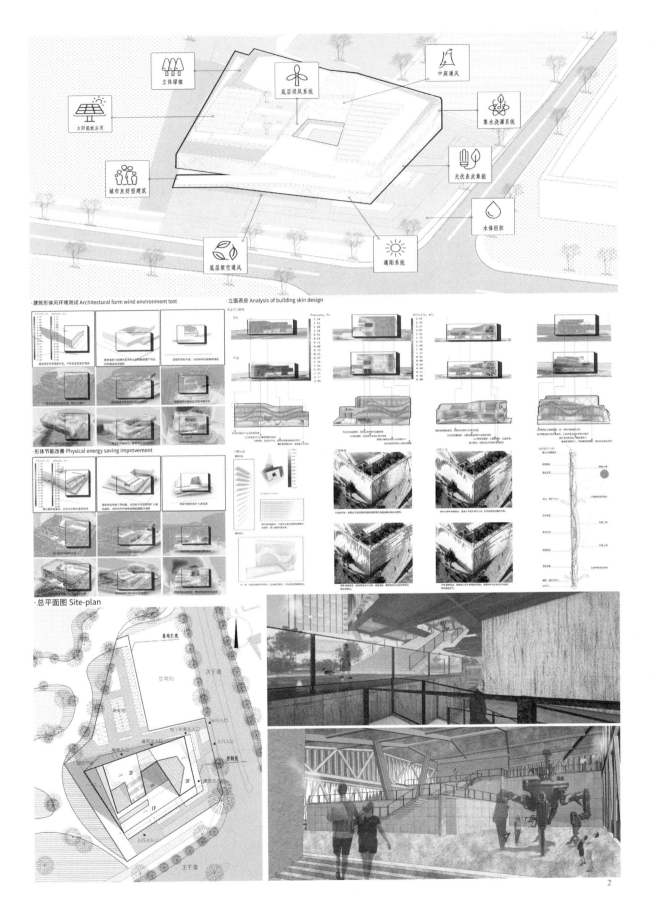

· 东南角透视图 Southeast Corner Perspective

· 首层平面图 Ground Floor Plan

· 平面爆炸图 Plane Explosion Pattern

8 交通盒　9 下沉广场
12 报告厅

· 绿色技术剖面图 Green Technology Profile

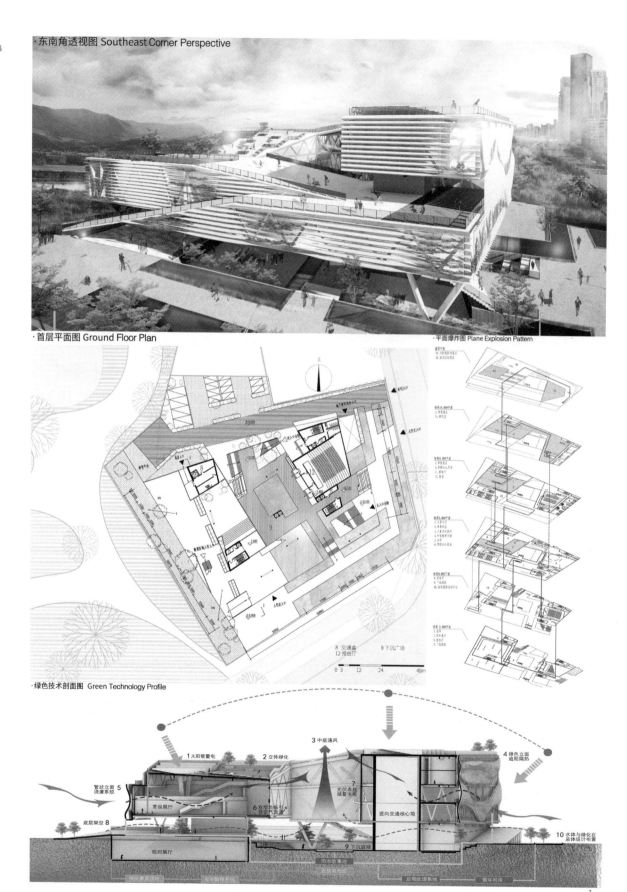

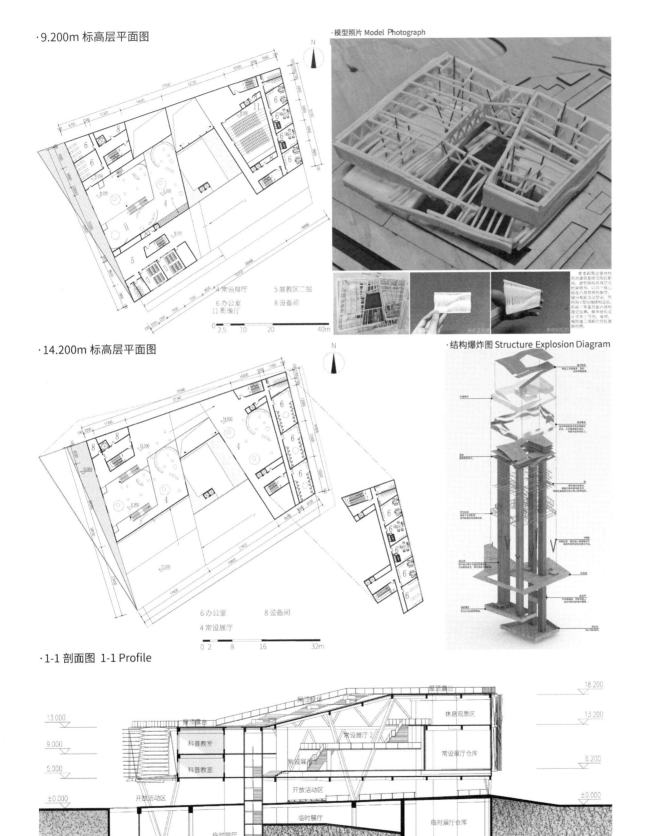

· 9.200m 标高层平面图

· 模型照片 Model Photograph

2023 台达杯国际太阳能建筑设计竞赛获奖作品集

4 常设展厅　　5 展教区二层
6 办公室　　　8 设备间
11 影像厅

· 14.200m 标高层平面图

· 结构爆炸图 Structure Explosion Diagram

6 办公室　　8 设备间
4 常设展厅

· 1-1 剖面图　1-1 Profile

综合奖・入围奖・零碳设计项目
Comprehensive Awards・Nomination・Zero-Carbon Design Project

注册号：101922
Register Number：101922

项目名称：编影・竹光
Entry Title：Weaving Shadow, Bamboo Light

作者：高静蔚、黎鑫、龙梦钰
Authors：Jingwei Gao, Xin Li and Mengyu Long

作者单位：南京工业大学
Authors from：Nanjing Tech University

指导教师：薛春霖、尤伟
Tutors：Chunlin Xue and Wei You

指导教师单位：南京工业大学
Tutors from：Nanjing Tech University

Design Description

科技馆位于增城区文化教育密集之处，将给周围的人群带来物质与精神上的双重影响。在设计中，我们运用了现代竹木技术，建筑主体用竹木胶合板制成，外侧遮阳板采用竹制编织技术，以现代工业化的方式实现量产。通过改变两种不同的"虚""实"可活动遮阳板的位置和角度，便于帮助建筑适应广州当地不同季节的气候。竹木材料契合了广州当地的气候条件，使得建筑内部自然舒适，且节能环保，在建筑建造与使用阶段起到良好的降碳作用。

The Science and Technology Museum is located in the dense cultural and educational area of Zengcheng District, which will bring both material and spiritual impacts to the surrounding population.

In the design, we used modern bamboo and wood technology, the main body of the building is made of bamboo and wood plywood, and the outer sunshade is made of bamboo weaving technology, which is mass-produced in a modern industrialized way. By changing the position and angle of the two different "virtual" and "real" movable sun shades, the building can be easily adapted to the different seasons of the local climate in Guangzhou. The bamboo and wood materials are adapted to the local climate conditions in Guangzhou, making the interior of the building natural and comfortable, as well as being energy efficient and environmentally friendly, reducing carbon emissions during the construction and use of the building.

Major Technical-Economic Indices

Site Area	11447㎡
Building Area	18429.4㎡
Floor Area Ratio	1.28
Building Density	26.1%
Building Height	24m

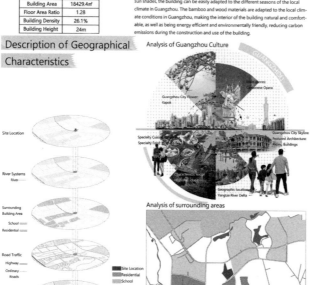

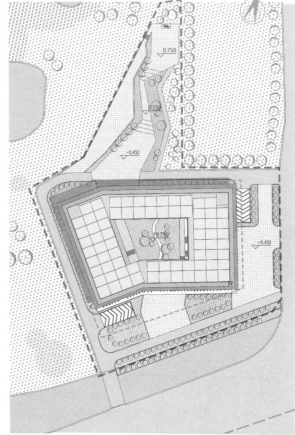

Site Plan 1：500

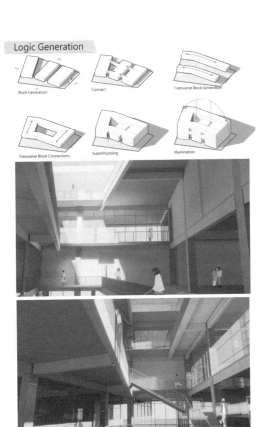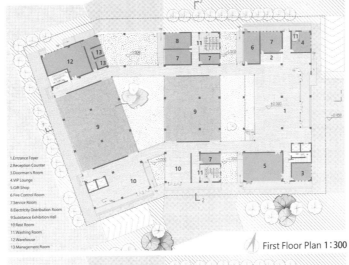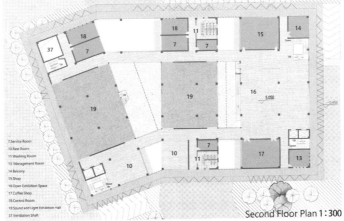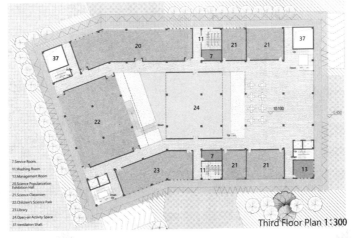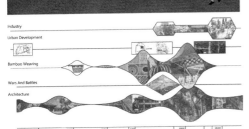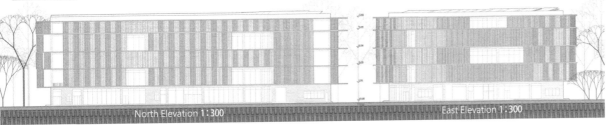

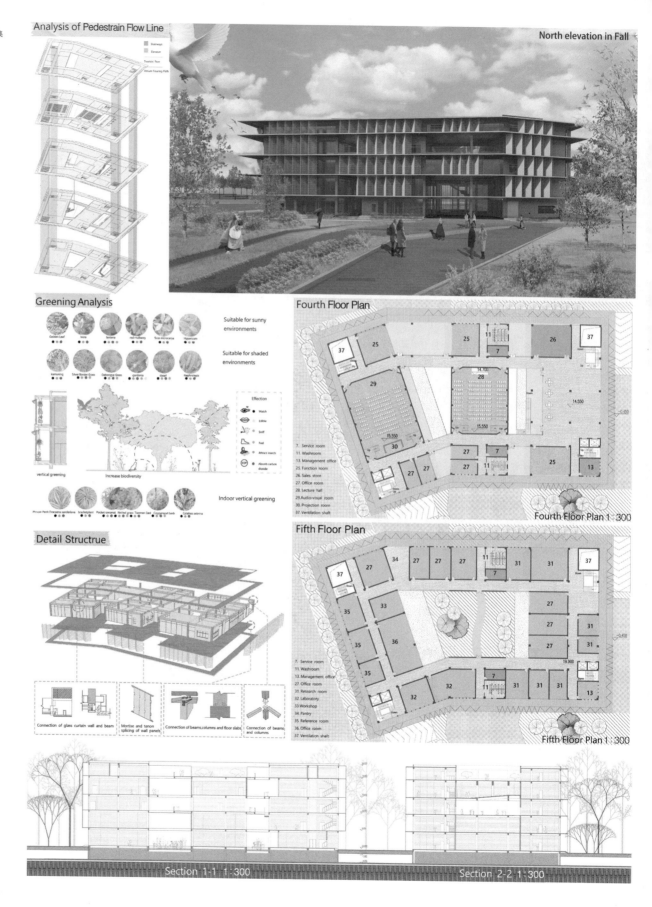

Energy Efficent Design

In terms of energy saving, we designed the building with both vertical and horizontal airflow exchange channels. In the longitudinal direction, we not only put in two large patios to facilitate airflow exchange between floors, but also added two small ventilation shafts to help local micro-climate regulation. In the horizontal direction, inspired by the cold alleys in traditional Lingnan architecture, we have placed a number of air ducts in the north-south and east-west directions to facilitate the exchange of airflow from room to room. In addition, we have left a channel for airflow exchange in the ceiling of each floor.

Rainwater collection

Descriptin of the Report

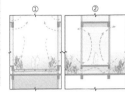

Note: 1. All the above calculations are based on floor area (public buildings)

Indoor Microclimate in Atrium

The atrium section has a comfortable airflow environment with relatively cool temperatures. The atrium also regulates airflow and air temperature throughout the building.

Site Climate Analysis

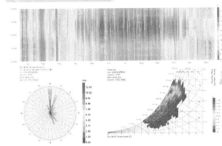

Airflow Analysis

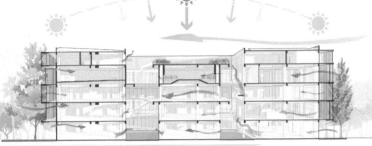

Analysis of Movable Sunshade for Airflow and Light Regulation

 By changing the position and angle of the sunshading panels, it is easy to help the building adapt to the different seasons of the local climate in Guangzhou and be more energy efficient.

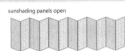

sunshading panels open | sunshading panels pull back

Transitional season (spring, fall)

Winter

Summer

(1) South side

Winds are more variable in the spring and fall. For movable screens, you can adjust the position and angle of the screen through the device.

The position and angle of the screen can be adjusted by the device to introduce airflow into the room and promote ventilation.

In winter, the dominant wind direction in this area is northwest, and the main effect is to bring cold airflow.

On the west and south sides, the screen can be closed to welcome the warmth of the sun.

In summer, the dominant wind direction in this area is southeasterly, and the main impact is to bring comfortable airflow.

On the west and south sides, the screens can be pulled back to shield the building from strong sunlight.

Sunlight Analysis

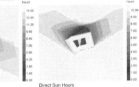

Direct Sun Hours | Direct Sun Hours

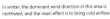

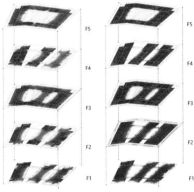

Summer | Winter

Radiance Analysis

Incident Radiation Summer | Incident Radiation Winter

Summer solar radiation

with sunshade | without sunshade

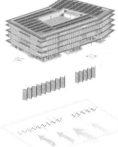

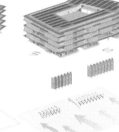

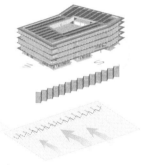

(2) North side

Timely adjustments can be made depending on the wind direction.

At this time, the north screen is pulled up, due to its special combination of real and imaginary, it can play the role of wind blocking and effectively prevent the cold airflow into the building interior, promote heat preservation.

At this time, the screen on the north side can be folded down, which effectively facilitates the flow of comfortable air into the interior of the building and enhances ventilation.

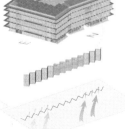

综合奖·入围奖·零碳设计项目
Comprehensive Awards · Nomination · Zero-Carbon Design Project

注册号：101982
Register Number：101982

项目名称：水境天梯
Entry Title：Waterfront Stairway

作者：王艺达、庄颖楠、杜明洋、张涵
Authors：Yida Wang, Yingnan Zhuang, Mingyang Du and Han Zhang

作者单位：重庆大学
Authors from：Chongqing University

指导教师：张海滨
Tutor：Haibin Zhang

指导教师单位：重庆大学
Tutor from：Chongqing University

本设计融合了独特的建筑设计构思、太阳能技术的创新运用以及低碳技术的智慧融合，呈现出一座富有新意的建筑。借鉴大自然的地理风貌，建筑倾斜如山，以湖泊为背景。设计不仅营造了令人印象深刻的视觉效果，还最大限度地优化了采光和通风，减轻了能源负担。建筑利用先进的太阳能技术，将太阳的能量转化为电力和热能，为建筑提供清洁能源。

太阳能板以创新的方式融入建筑外观，不仅减少了碳排放，还为周边社区提供了可再生能源。建筑内部采用了最新的低碳技术，如智能照明、节能设备和可持续材料。此外，高耸中庭和开放式设计鼓励自然通风，最大限度地降低了对人工空调的依赖。设计创新为科教之心注入了新的生命和活力。本设计旨在探索可持续发展，结合建筑设计的美感和高效能源利用，为未来的科学、教育和文化活动提供了一片可持续的天地。

This design incorporates a unique architectural concept, innovative use of solar technology, and intelligent integration of low-carbon techniques, presenting a remarkably innovative building. Drawing inspiration from nature's graceful slopes, the architecture tilts like a mountain against the backdrop of a serene lake. This design not only creates an impressive visual effect but also optimizes natural lighting and ventilation to the fullest extent, reducing the energy burden. The building harnesses advanced solar technology, converting solar energy into electricity and heat, providing clean energy to the structure. Solar panels are seamlessly integrated into the building's exterior in an innovative manner, reducing carbon emissions and supplying renewable energy to the surrounding community.

Internally, the building incorporates cutting-edge low-carbon technologies such as smart lighting, energy-efficient appliances, and sustainable materials. Additionally, the soaring atrium and open design encourage natural ventilation, significantly reducing reliance on artificial air conditioning. This design innovation injects new life and vitality into the Center for Science and Education. This design aims to explore sustainability by combining aesthetic architectural design with efficient energy utilization, providing a sustainable environment for future scientific, educational, and cultural activities.

Site Analysis

The site is located in the central axis and comprehensive functional area of Guangzhou Science and Education City in Zengcheng District, Guangzhou, at the intersection of Yushui 1st Road and Science and Education Avenue. East longitude 113.67°, North latitude 23.29°. Zengcheng District is located in the coastal zone near the east of Guangzhou, with a humid, hot and rainy climate throughout the year. The total land area is 11,447 square meters. A 110kV substation is planned to be built on the northeast side of the site.

Surrounding　Nature Resource　Transportation

Climatic Analysis

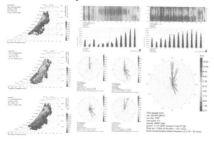

Site Plan 1:1000

Economic & Technical Indicators:

Site area: 11447 m²　Floor space: 4598 m²
Green ratio: 0.14　Building density: 40.2%
Plot ratio: 1.65

Road square: 5201 m²
Green area: 1610 m²
Ground floor: 18885.4 m²
Above-ground area: 15185.4 m²
Underground area: 3700 m²

Exhibition & education: 9997.4 m²
Communal service: 2460 m²
Business research room: 1362 m²

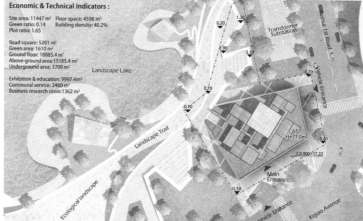

Concept

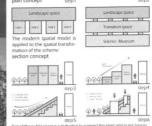

The modern spatial model is applied to the spatial transformation of the scheme section concept

Part of the public space is subdivided to connect the inner and outer spaces in a transitional space. Adjust the space size according to the demand, use the inclined plane to strengthen the space transformation logic, arrange the air shaft overhead layer to achieve ventilation.

Form Generation

| 01 | Site Information | 02 | Place volumes | 03 | Divide the mesa |
| 04 | Sloped roof | 05 | Insert a block | 06 | Roof boardwalk |

1st Floor Plan

1. lobby & Temporary exhibition hall
2. Security check
3. reception
4. toilet
5. office
6. coffee
7. souvenir shop
8. Multi-Function Hall
9. Secondary foyer
10. Storage room
11. Logistics foyer

Streamline

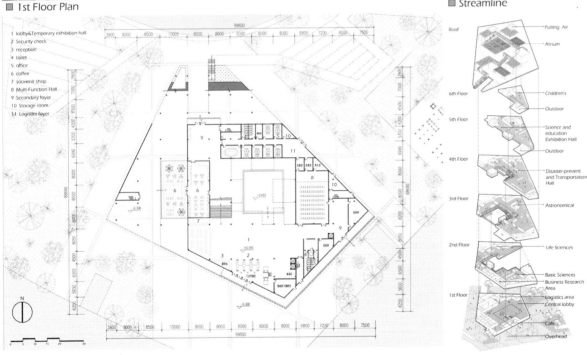

Perspective Section 1-1

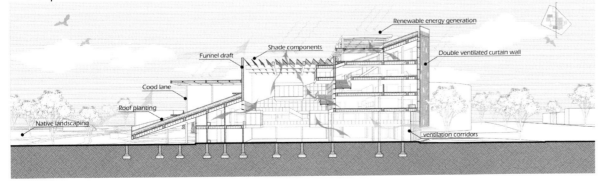

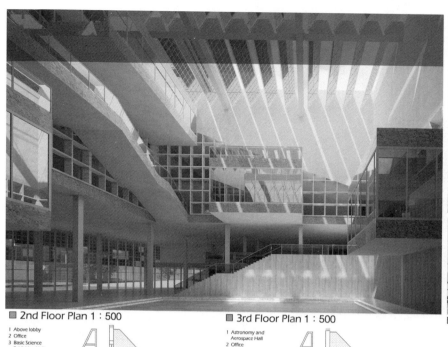

2nd Floor Plan 1:500
1 Above lobby
2 Office
3 Basic Science Exhibition Hall
4 Dubbing
5 Energy Exhibition Hall
6 Life Science & Technology Hall
7 Activity Room
8 Interactive area

3rd Floor Plan 1:500
1 Astronomy and Aerospace Hall
2 Office
3 Astronomical VR Experience Hall
4 Terrace
5 Eight Planetary Models
6 Spacewalk
7 Globes
8 Great court

4th Floor Plan 1:500
1 Astronomy and Aerospace Hall
2 Featured Exhibition Hall
3 Earthquake simulation room
4 Sunshade louver
5 Fire extinguishing experience area
6 High-speed train
7 Simulated driving
8 Great court
9 Jiaolong model

5th Floor Plan 1:500
1 Building Energy Efficiency Exhibition Hall
2 Over the exhibition hall
3 Energy saving process simulation exhibition hall
4 Sunshade louver
5 Interactive screen
6 Building Scale Model
7 Model booth
8 Adjustable photovoltaic glass
9 Terrace

Active Solar Technology

Rooftop solar panels generate electricity and provide shade. On the east, south and north facades, solar panels are installed in the facade frame, and its back is made of aluminum, which can be rotated as required; The west is framed with green shading.

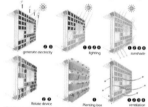

Optoelectronic system setup

Renewable energy capacity (use phase)

Energy consumption calculation results

Carbon Footprint

This building proactively incorporates a range of cutting-edge low-carbon technologies to reduce its carbon footprint. Firstly, it utilizes a solar photovoltaic system that converts solar energy into electricity, providing clean energy to the building while significantly reducing traditional energy consumption and carbon emissions. Secondly, an intelligent lighting system automatically adjusts brightness and energy usage, minimizing energy wastage. Additionally, high-efficiency insulation materials and ventilation systems are employed to optimize the building's energy efficiency, further lowering its carbon footprint.

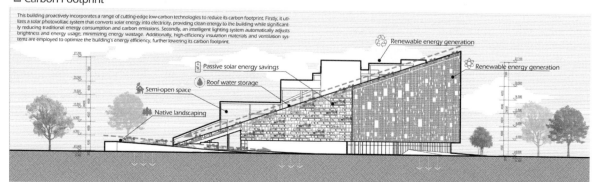

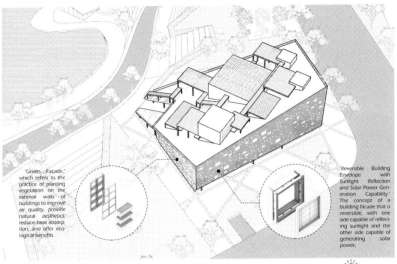
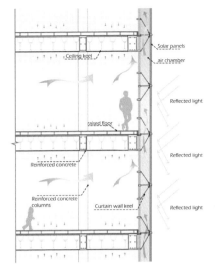

"Green Facade," which refers to the practice of planting vegetation on the exterior walls of buildings to improve air quality, provide natural aesthetics, reduce heat absorption, and offer ecological benefits.

"Reversible Building Envelope with Sunlight Reflection and Solar Power Generation Capability." The concept of a building facade that is reversible, with one side capable of reflecting sunlight and the other side capable of generating solar power.

■ Ventilation Strategies

In summer, the bottom layer is elevated, and the overhead layer is often open. The Venturi tube phenomenon is used to open the wall arrangement, and the wind speed is slow in the wide area and fast in the narrow area to accelerate the bottom air flow.

In winter, the bottom layer is overhead and the overhead layer is usually closed, which reduces the wind entering the room and reduces the heat exchange. At the same time, the overhead layer serves as the transition between indoor and outdoor space and reduces the transfer efficiency

Through the skylight in the atrium, heat exchange is realized. The heating of the air at the top causes the air at the bottom to rise, and through the air flow, the hot air in the exhibition space on each floor flows into the atrium

The overhead floor and skylight are closed to reduce heat loss, and the air in the atrium is heated to carry out thermal circulation, driving the air flow of each layer

■ Carbon Emissions

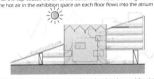

Green building review of energy conservation and energy utilization

■ Radiation Simulation

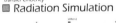

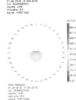
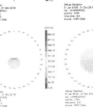
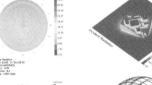

■ Wind Simulation

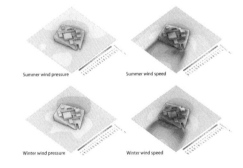

■ Plant Footprint

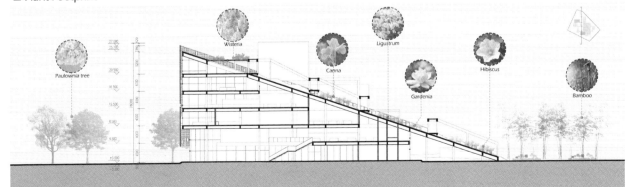

综合奖・入围奖・零碳设计项目
Comprehensive Awards・Nomination・Zero-Carbon Design Project

注册号：102012
Register Number：102012

项目名称：风穿绿廊 碳循未来
Entry Title：Wind Through Green Gallery, Carbon Recycling for the Future

作者：赵庆卓、郭奕岑、张康睿、陈忠耀
Authors：Qingzhuo Zhao, Yicen Guo, Kangrui Zhang and Zhongyao Chen

作者单位：重庆大学
Authors from：Chongqing University

指导教师：周铁军、张海滨
Tutors：Tiejun Zhou and Haibin Zhang

指导教师单位：重庆大学
Tutors from：Chongqing University

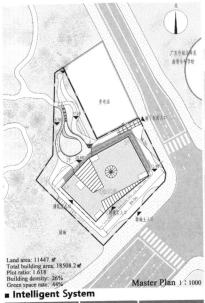

Land area: 11447 ㎡
Total building area: 18508.2 ㎡
Plot ratio: 1.618
Building density: 26%
Green space rate: 44%

Master Plan 1:1000

■ Intelligent System

Green & Energy Saving	Safety & Security	Efficient & Convenient
Intelligent air conditioning system	surveillance system	Indoor air quality detection and control system
Intelligent power monitoring system	Personnel and vehicle control systems	
resource recycling system	fire detection system	Indoor lighting detection and control system
Intelligent lighting system	disaster response system	
Building carbon emission monitoring system		

■ Volume Generation

01.Surroundings
The land is divided into building land and landscape land, and a circular driveway is set to surround the building

02.Functional partition
Simply divide the main body of the building into two volumes: the logistics part and the exhibition hall part

03.The first floor
Merge the ground floor of the building and connect the building with the north side landscape

04.Aerial corridor
An aerial corridor connects the top of the building and forms part of the exhibition hall

05.Entrance
The building volume is cut inward to strengthen the entrance form and guide people towards the stairs

06.Impracticable
The first and second floors of the building are raised to achieve better spatial experience and wind environment

07.Wind corridor
The facade of the building is recessed to highlight the location of the wind corridor

08.Roof
Glass shading is made on the roof of the building to indicate the building volume, and photovoltaic panels are laid according to the calculation

■ Aerial View

■ Outdoor Rendering

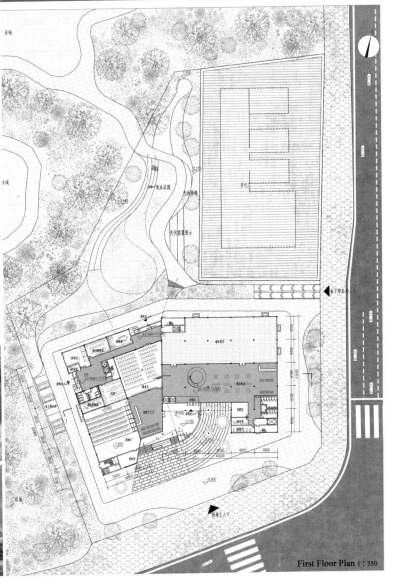

First Floor Plan 1:350

■ Technical Details Analysis

Mezzanine Floor Plan 1:350   Second Floor Plan 1:350

■ Scene Rendering

■ Exploded Axonometric & Materials

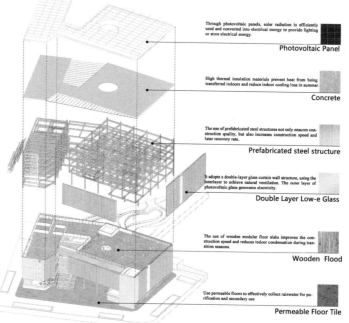

Through photovoltaic panels, solar radiation is efficiently used and converted into electrical energy to provide lighting or store electrical energy.
Photovoltaic Panel

High thermal insulation materials prevent heat from being transferred indoors and reduce indoor cooling loss in summer.
Concrete

The use of prefabricated steel structures not only ensures construction quality, but also increases construction speed and later recovery rate.
Prefabricated steel structure

It adopts a double-layer glass curtain wall structure, using the interlayer to achieve natural ventilation. The outer layer of photovoltaic glass generates electricity.
Double Layer Low-e Glass

The use of wooden modular floor slabs improves the construction speed and reduces indoor condensation during transition seasons.
Wooden Flood

Use permeable floors to effectively collect rainwater for purification and secondary use
Permeable Floor Tile

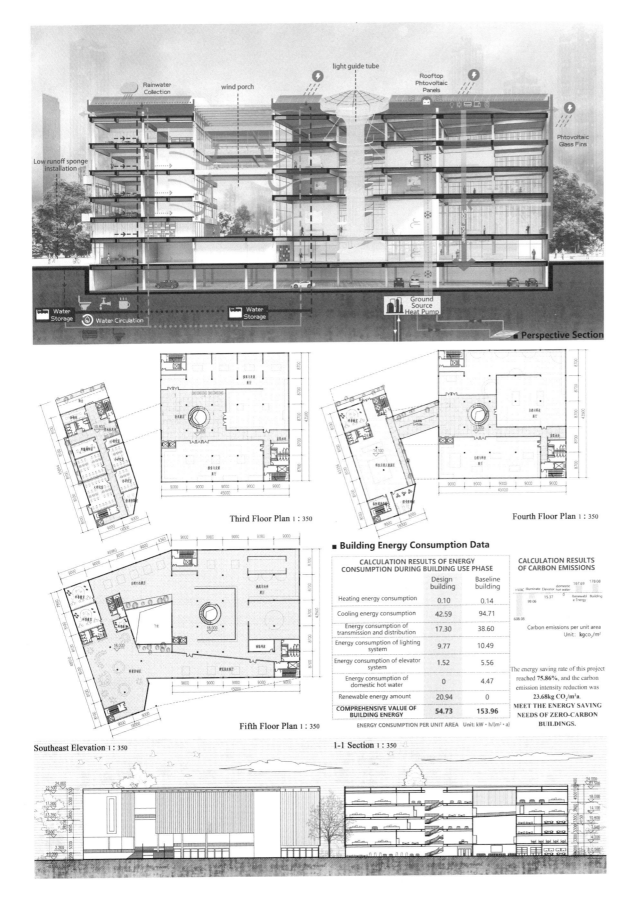

综合奖・入围奖・零碳设计项目
Comprehensive Awards · Nomination · Zero-Carbon Design Project

注册号：102013
Register Number：102013

项目名称：风起·光旋
Entry Title：Wind Swirling with Light

作者：聂乐仪、王依然、王晗、韦茹钰、唐一文
Tutors：Leyi Nie, Yiran Wang, Han Wang, Ruyu Wei and Yiwen Tang

作者单位：重庆大学
Authors from：Chongqing University

指导教师：周铁军、张海滨
Tutors：Tiejun Zhou and Haibin Zhang

指导教师单位：重庆大学
Tutors from：Chongqing University

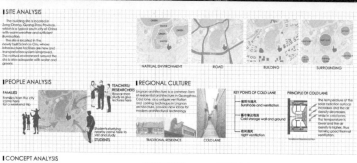

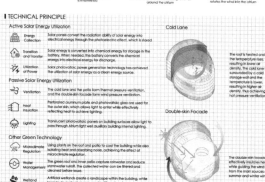
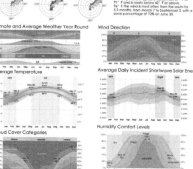

风起·光旋
Wind Swirling with Light 2

Exhibition hall

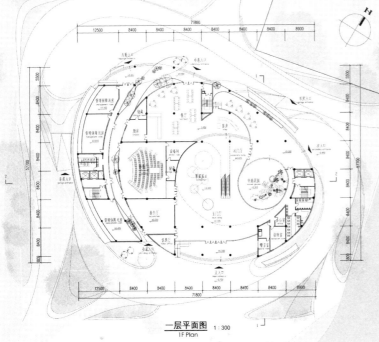

一层平面图 1:300
1F Plan

Secondary entrance

SOLAR PHOTOVOLAIC SYSTEM

Technical Strategy

Photovoltaic flexible film hollow louver curtain wall
Perforated aluminum plate covered with photovoltaic flexible film

Technical Route
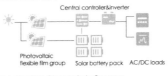
Central controller & inverter
Photovoltaic flexible film group   Solar battery pack   AC/DC loads

Exterior Skin Photovoltaic Strategy
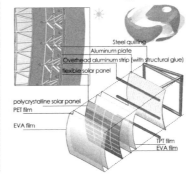
Steel quilting
Aluminum plate
Overhead aluminum strip (with structural glue)
flexible solar panel
polycrystalline solar panel
PET film
EVA film
TPT film
EVA film

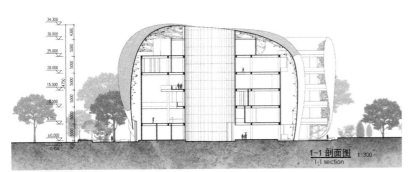

1-1 剖面图 1:300
1-1 section

Curtain Wall Photovoltaic Strategy
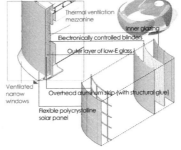
Thermal ventilation mezzanine
Inner glazing
Electronically controlled blinders
Outer layer of low-E glass
Ventilated narrow windows
Overhead aluminum strip (with structural glue)
Flexible polycrystalline solar panel

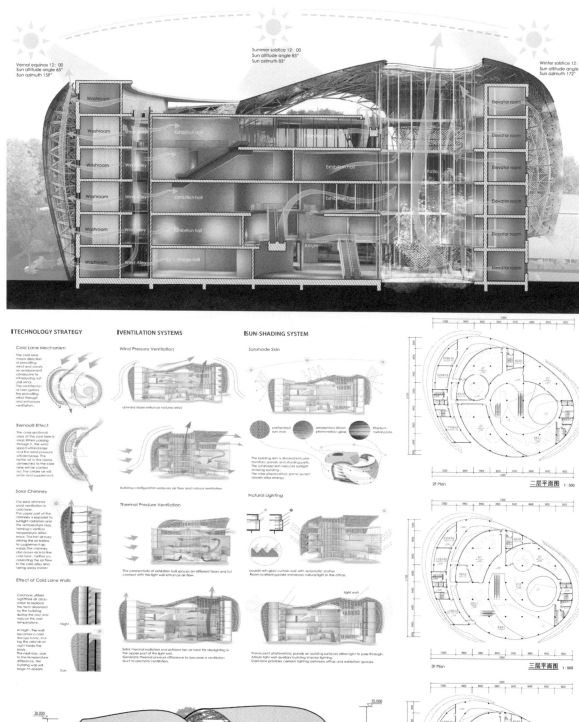
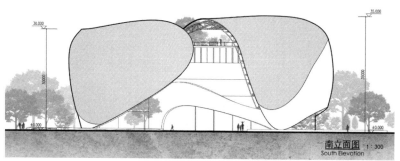
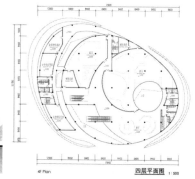

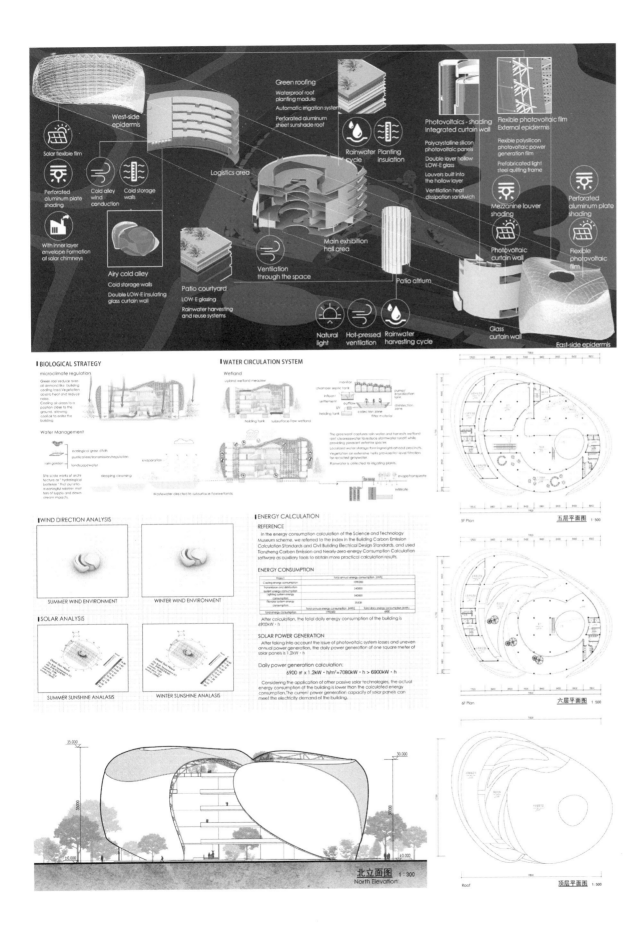

综合奖·入围奖·零碳提升项目
Comprehensive Awards · Nomination · Zero-Carbon Promotion Project

注册号：101321
Register Number：101321

项目名称：叶园
Entry Title：The Vein

作者：刘为群、孙艺玮
Authors：Weiqun Liu and Yiwei Sun

作者单位：厦门大学
Authors from：Xiamen University

指导教师：张燕来
Tutor：Yanlai Zhang

指导教师单位：厦门大学
Tutor from：Xiamen University

场地区位

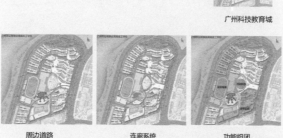

设计场地位于广州市增城区朱村街道广州科技教育城内，距离广州市中心城区45千米，包括13所市属职业院校(含技校)、交通及市政配套设施，三大组团共享带、安置区、四大公园等，可容纳学生约12.9万人，欲将其打造成具有岭南特色的山水田园型教育城。

区位气候

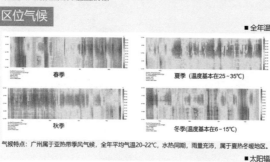

气候特点：广州属于亚热带季风气候，全年平均气温20~22℃，水热同期，雨量充沛，属于夏热冬暖地区。

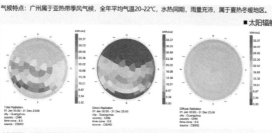

太阳辐射：北回归线穿越广东中部，太阳高度角大，日照时间长且辐射量大。

自然通风潜力分析

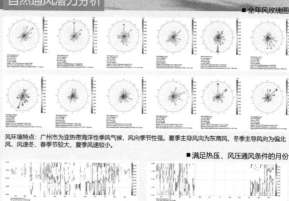

风环境特点：广州市为亚热带海洋性季风气候，风向季节性强。夏季主导风向为东南风，冬季主导风向为偏北风，风速冬、春季节较大，夏季风速较小。

从上图可看出，广州地区全年基本可满足热压通风条件，而由于夏秋季节风速较低，只有冬、春季节可满足风压通风条件。后续设计可以考虑多采用热压通风措施来对建筑自然通风进行改善。

从上图可看出，满足热压通风条件的主要是来自东北向的风。

该社团中心受周边建筑遮挡较大，整体风速小于0.94m/s，北侧体块通风较好，东侧为主迎风面。

总体规划

整体鸟瞰图　　场地鸟瞰图

广州科技教育城以现状山水为基底，建设连山通水的绿色网络，以职业教育为特点，组织城校融合的功能布局。以岭南风格为指引，打造地域特色的形象设计。广州科技教育城入驻13所院校，容纳学生12.9万人。通过保留自然地貌特征，建立城市与山体之间的视线通廊，重点打造"四大公园、一条水系、四季景观"，凸显"花城、绿城、水城"特色。

Based on the current landscape, Guangzhou Science and Technology Education City builds a green network connecting mountains and rivers, features vocational education, and organizes the functional layout of city and school integration. Lingnan style as a guide to create regional characteristics of the image design. Guangzhou Science and Technology Education City has 13 colleges and universities, accommodating 129,000 students. By retaining natural landform features, the establishment of a line of sight corridor between the city and the mountain, focusing on creating "four parks, a water system, four seasons landscape", highlighting the characteristics of "flower city, green city, water city".

规划理念

山水　以现状山水为基底，建设连山通水的绿色网络

利用现状西福河、灌溉渠、山塘和鱼塘，规划4条联通水系、2个景观湖、1个漫游公园，形成"上部山塘蓄水、中部截洪沟导水、下部湿地公园净水"的水文安全格局，以白水山为生态核心构建三条联系北部山体和西南部河流的生态廊道。

低碳　以低碳生态为核心，打造绿色智慧的支撑系统

规划设置了"资源合理利用""环境质量良好""经济持续发展""社会和谐进步"4个目标、32个分项的指标体系，并将绿色生态指标落实到控制管理单元，同时通过绿色交通、绿色建筑、绿色市政等专项，确保绿色生态城区的开发建设可量化、能落地、看得见。

岭南　以现状山水为基底，建设连山通水的绿色网络

通过对通风、噪声、日照等要素模拟，设计符合岭南气候特征的建筑空间布局。结合综合交通枢纽形成南高北低的空间形态，展现从城市到生态的演进。

概念生成

 +  =

以叶片脉络为原型　　遵照项目需求规划功能　　连接岛屿之间构成环路　　叠加形成横向与纵向交织
提取其肌理为交通网络。　顾应建筑划分若干岛屿。　　构建公共空间和节点广场。　有机自然的建筑系统。
Connecting greenlandscape corridors　Function Islands　Public Loop with spaces & plazas　Form an organic and natural system

设计元素

 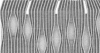 

编织立面　　拉开幕帘　　引入绿洲
Woven façade　Pull the curtain　Cultivate the Oasis

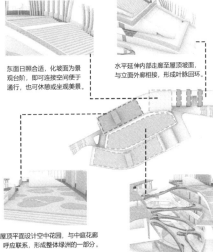

东面日照合适，化坡面为景观台阶，即可连接空间便于通行，也可休憩或坐观美景。

水平延伸内部走廊至屋顶坡面，与立面外廊相接，形成叶脉回环。

屋顶平面设计空中花园，与中庭花廊呼应联系，形成整体绿洲的一部分。

中庭连廊错落交叠，如叶脉的编织与撕裂，将活力的营养液输送到教学楼的每一处。

节能策略

屋顶平台花园 — 北侧的屋顶空间相比于南侧所受到的太阳辐射量较小,因此决定将北侧改造为露天的开放花园,称之为"暖园";而南侧则是有所遮盖的、可避风遮雨的休憩空间,称之为"凉园"。

中庭步道连廊 — 中庭的连廊作为交通空间编织的核心交接处,将橙色作为其主要色彩元素,不仅仅是一种视觉吸引力,还代表着热情、活力和创造力。橙花廊象征着学生在这个空间中的成长与发展,激励着他们勇于探索和创新。

北侧受太阳辐射较弱

南侧受太阳辐射较强

太阳能光伏系统
太阳能集热屋面墙体

墙体编织的垂直交叉的线条如幕帘般拉开,从中可以窥见于教学楼基座、景观花廊以及屋顶上廊的绿洲,让人感觉幕帘里藏着一个"花海世界"。

通透植被构架

在通风进入的风道上加入绿植或者水面等元素,降低进入室内空气的温度。风吹过水面之后会带有一定的湿度,从而保证了新鲜空气的供应。

蒸发冷却系统

绿化降温净化 — 设计立体绿化,将生态性、人性化的因素融入到建筑之中,在建筑的外部空间进行绿化,设置花架、连廊等以减少日辐射对周围环境的影响,改善小气候。

自然通风

室内采光

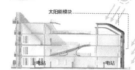
太阳能光伏发电

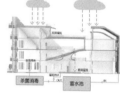
雨水收集

建筑使用阶段碳排放计算

项目		设计建筑		基准建筑	
		碳排放量(kgCO₂)	单位面积碳排放(kgCO₂/m²)	碳排放量(kgCO₂)	单位面积碳排放(kgCO₂/m²)
碳排放项	暖通空调	1084656.05	319.02	2346032.19	938.41
	生活热水	0.00	0.00	0.00	0.00
	照明	484610.64	142.53	727490.56	291.9
	电梯	46521.30	13.68	54917.94	21.97
减碳项	可再生能源	1856042.0	545.89	0.00	0.00
	建筑碳汇	457860	134.66	0.00	0.00
合计		-698.11	-0.21	3128441.69	1251.38

改造后内廊的露天部分将成为学生和教职工之间的社交和学术互动场所。它可以用于展览、演讲、艺术展示、小型聚会等各种活动,为建筑带来了更多的功能。本设计将建筑与自然相融合,为建筑创造一个有生命力的内部空间。

经济技术指标：
占地面积：1700m²
建筑面积：3400m²
建筑高度：15.45m
建筑层数：4层

2023 台达杯国际太阳能建筑设计竞赛获奖作品集

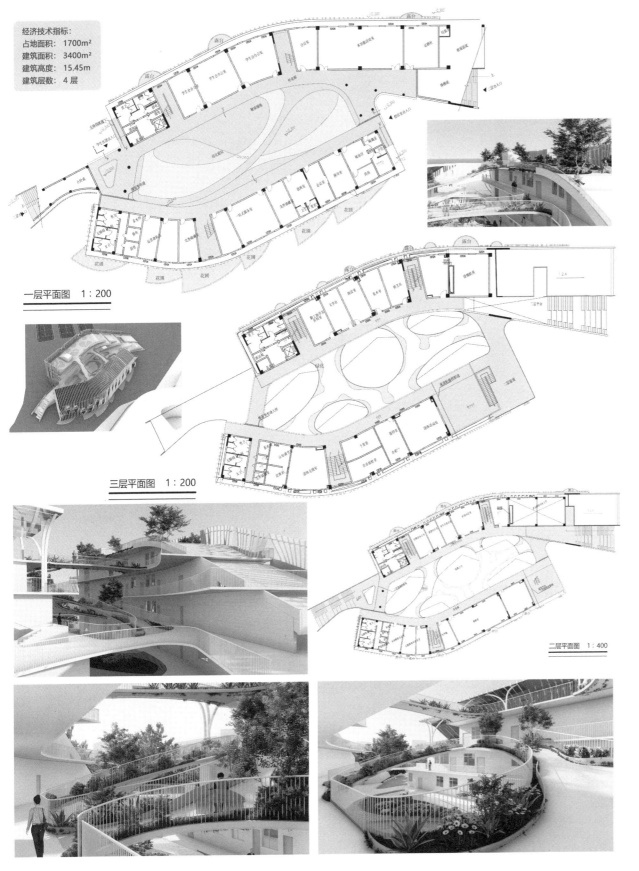

一层平面图 1：200

三层平面图 1：200

二层平面图 1：400

综合奖·入围奖·零碳提升项目
Comprehensive Awards · Nomination · Zero-Carbon Promotion Project

注册号：101406
Register Number：101406

项目名称：积木拼荫
Entry Title：Brick Puzzled Shadows

作者：卓越、赵钧彦、邓苏媛、阮景添、李梓杰
Authors：Yue Zhuo, Junyan Zhao, Suyuan Deng, Jingtian Ruan and Zijie Li

作者单位：福州大学
Authors from：Fuzhou University

指导教师：邱文明
Tutor：Wenming Qiu

指导教师单位：福州大学
Tutor from：Fuzhou University

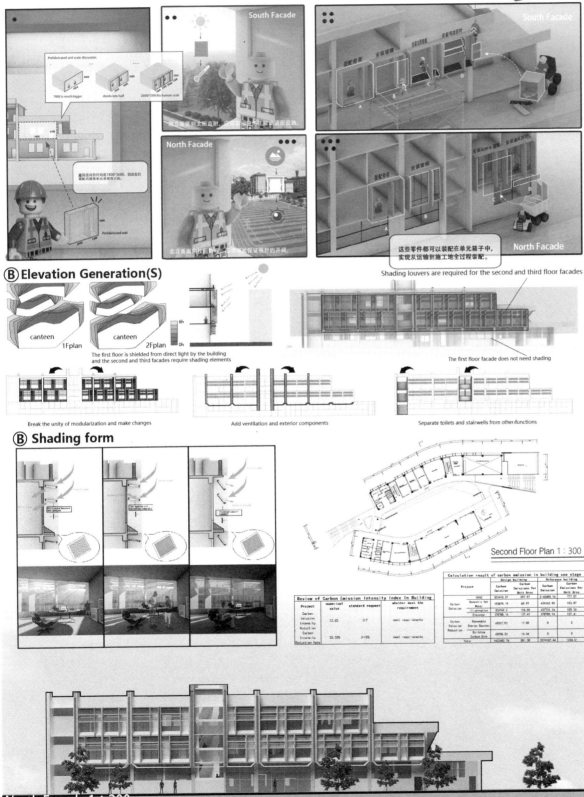

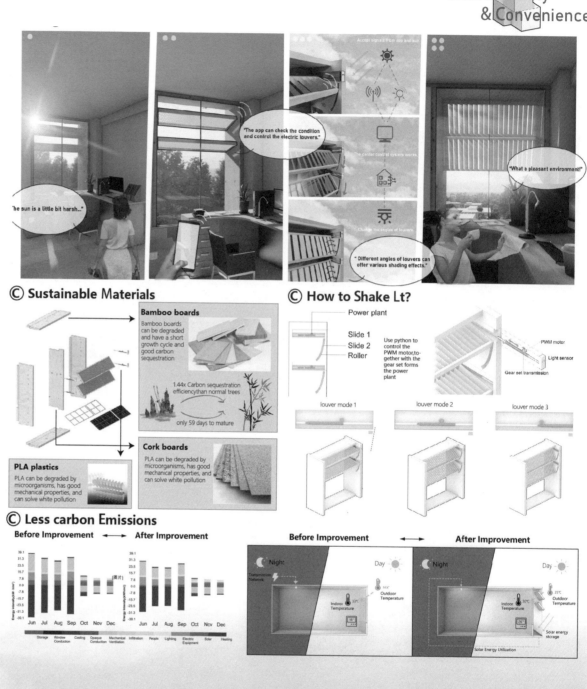

Green technology

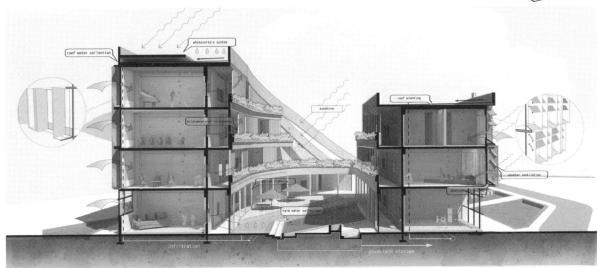

Ⓓ Solar Energy Technology

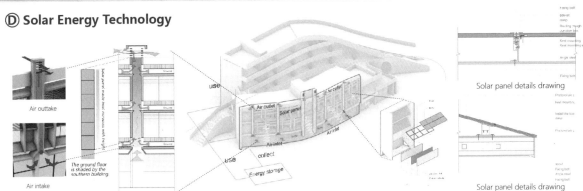

Ⓓ Rainwater Collection and Roof Greening

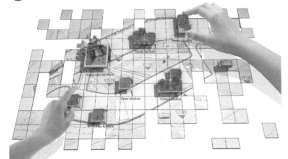

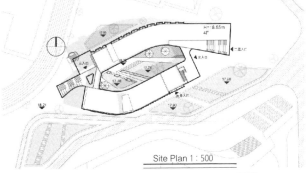

Site Plan 1 : 500

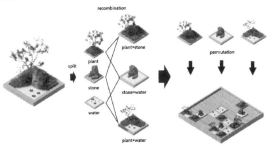

plant units details drawing

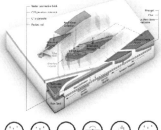

综合奖·入围奖·零碳提升项目
Comprehensive Awards · Nomination · Zero-Carbon Promotion Project

注册号：101528
Register Number：101528

项目名称：零碳方舟，"碳"未来
Entry Title：Zero-Carbon Ark，"Carbon" Surfing the Future

作者：康孟琪、赵琳、吕铭洁、姚望、王君宜
Authors：Mengqi Kang, Lin Zhao, Mingjie Lyu, Wang Yao, and Junyi Wang

作者单位：南京工业大学
Authors from：Nanjing Tech University

指导教师：胡振宇、张海燕
Tutors：Zhenyu Hu and Haiyan Zhang

指导教师单位：南京工业大学
Tutors from：Nanjing Tech University

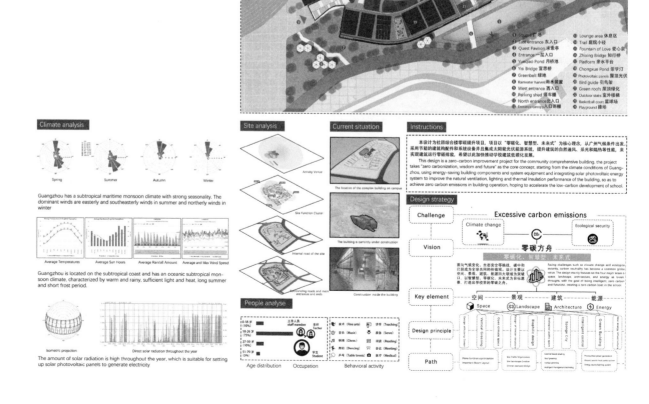

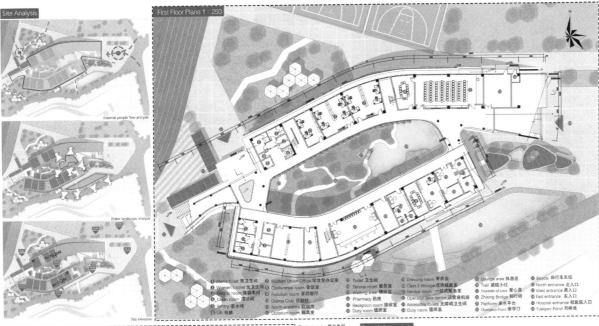
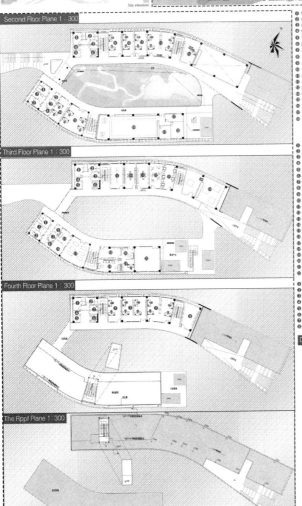

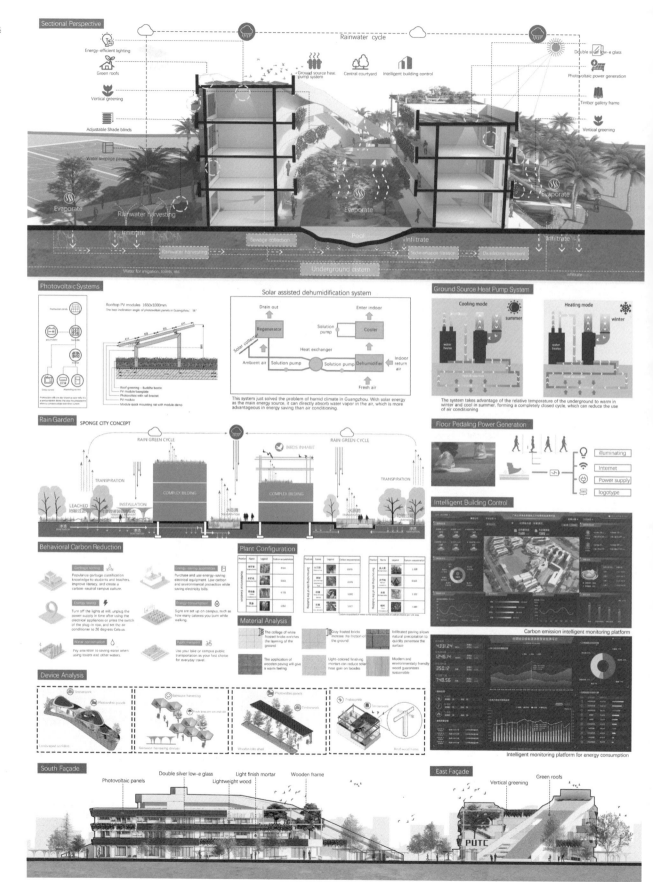

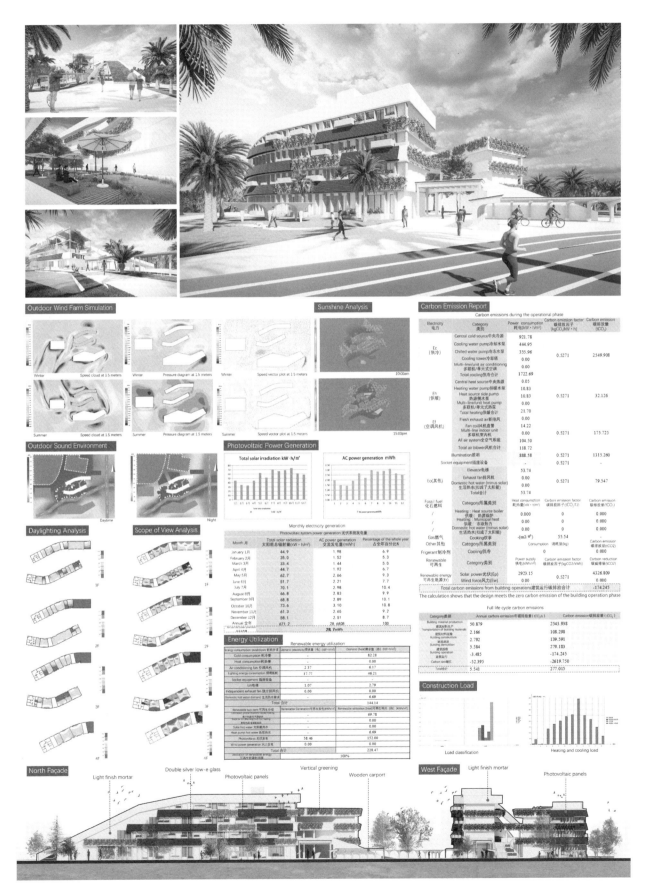

综合奖・入围奖・零碳提升项目
Comprehensive Awards・Nomination・Zero-Carbon Promotion Project

注册号：101714
Register Number：101714

项目名称：既改＋
Entry Title：All in One

作者：赵翔宇、张冉晨、史国崇、庞志祥、张春杰
Authors：Xiangyu Zhao, Ranchen Zhang, Guochong Shi, Zhixiang Pang and Chunjie Zhang

作者单位：大连理工大学
Authors from：Dalian University of Technology

指导教师：李国鹏
Tutor：Guopeng Li

指导教师单位：大连理工大学
Tutor from：Dalian University of Technology

The project is located in Guangzhou City, Guangdong Province, China. According to the thermal zoning for Chinese buildings, it falls into the hot summer and warm winter zone, specifically in the southern region. The architectural design must thoroughly consider summer heat prevention due to the hot summers in this area.
Guangzhou City experiences high outdoor humidity, with an annual average relative humidity of 76%. The average humidity is particularly high from June to September during the summer. The total annual solar radiation can accumulate to 4090MJ/m², with significant radiation during the summer.
The outdoor wind speed is mainly concentrated between 1-1.5m/s, 2-2.5m/s, and 3-3.5m/s. Given Guangzhou's monsoonal climate, wind directions vary seasonally, with the most frequent winds coming from the northeast and the least from the northwest.

设计说明：该设计从既有建筑基础条件出发，找出既有建筑不利于节能减碳的缺点，以被动优先、主动优化的原则进行改造；通过对建筑主体加装遮阳构件、改变窗户形式及位置、设置气候中庭、增加雨水回收装置等手段以达到节约能源，降低碳排的目的，同时充分新能源——太阳能，对建筑进行能源补充。屋面采用创新的设计和布置方式，达到更高的产能效益，立面选用发电玻璃，满足遮阳条件的同时还能够辅助发电；最后以节能设备和智能控制进行辅助，降低建筑运行期间的能耗，以系统化的设计手段在节能减碳的同时保障建筑使用时的舒适性。

Design Statement: The design starts from the basic conditions of the existing buildings, identifies the shortcomings of the existing buildings that are not conducive to energy saving and carbon reduction, and carries out the renovation based on the principle of passive priority and active optimization: through the addition of shading components to the main body of the building, changing the form and location of the windows, setting up a climatic atrium, and increasing the rainwater recycling device to achieve the purpose of energy saving and carbon emission reduction, and at the same time, fully utilizing the new energy source - solar energy - to supplement the energy of the building. solar energy to supplement the building's energy, adopting an innovative east-west arrangement for the roof to achieve higher energy efficiency, and choosing power generation glass for the facade to meet the sun-shading conditions while also assisting in power generation; and finally assisting with energy-saving equipments and intelligent control to reduce energy consumption during the operation of the building, so as to guarantee the comfort of the building when using it while saving energy and reducing carbon emissions by means of systematized design methods. The systematic means to reduce energy consumption during the operation of the building, and at the same time ensure the comfort of the building.

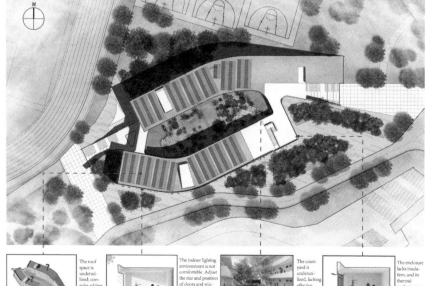

被动式太阳能利用 Passive Solar Energy Utilization

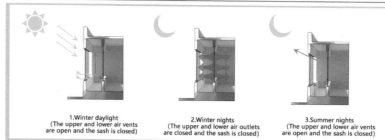

1. Winter daylight
(The upper and lower air vents are open and the sash is closed)

2. Winter nights
(The upper and lower air outlets are closed and the sash is closed)

3. Summer nights
(The upper and lower air vents are open and the sash is closed)

太阳能供电的微喷灌系统 Solar-powered micro-sprinkler irrigation system

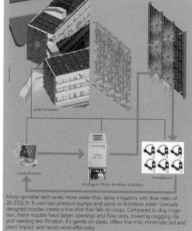

Micro-sprinkler tech saves more water than spray irrigation, with flow rates of 20-250L/h. It uses low-pressure pumps and pipes to distribute water. Specially designed nozzles create a fine mist that falls on crops. Compared to drip irrigation, these nozzles have larger openings and flow rates, lowering clogging risk and needing less filtration. It's gentle on pipes, offers fine mist, minimizes soil and plant impact, and resists wind effectively.

太阳能光热 Solar Thermal

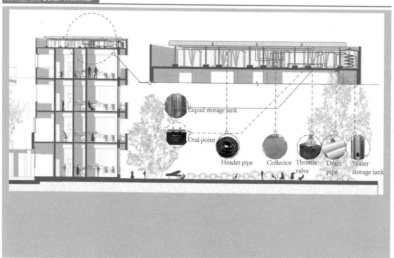

庭院太阳能灯具 Garden Solar Lamps and Lanterns

Using warm 3000K lighting creates a cozy night ambiance. Smart lighting highlights building and tree outlines. Solar lights in the courtyard save on power.

屋面光伏发电 Rooftop Photovoltaics

太阳能光电 Solar Photovoltaics

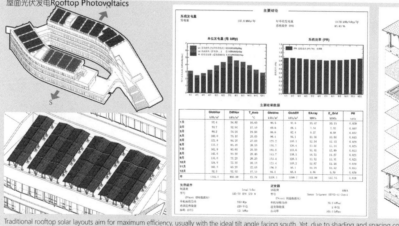

Traditional rooftop solar layouts aim for maximum efficiency, usually with the ideal tilt angle facing south. Yet, due to shading and spacing constraints, this layout may not maximize electricity per unit area. **As solar panel costs drop, optimizing layouts by increasing panel numbers and reducing spacing boosts overall electricity in a given area.** In this design, we propose an innovative solar panel layout. Simulation indicates a 12° tilt and east-west orientation generate maximum electricity within the set area while allowing for maintenance. With this, [number of panels] solar panels can be installed. In Guangzhou, this yields an annual total of 137800kW·h and a daily average of about 378kW·h, significantly offsetting the building's energy use.

随着光伏板成本的下降，可以采用更加优化的光伏板排布形式，通过增加光伏板数量，减小光伏板间距来增加单位面积内的发电总量。此次建筑设计中我们提出创新光伏板排布方式，经过模拟与计算，光伏板倾角为12°、东西朝向排布时，可以在规定安装面积内产生最大发电量，且留有足够的检修空间。

立面发电玻璃组件 Facade Power Generation Glass Module

Using 314 photovoltaic glass panels (1.8×0.85=1.53sq. meters each), generating 200-210kW·h annually per panel. About 25%~30% of panels have reduced efficiency due to shading adjustments, halving their output. Considering the maximum reduction for 94 panels (30%), they generate an estimated 9400kW·h annually (94×200×0.5). Hence, conservatively, the total annual electricity from these panels is around 53,400kW·h (44,000kW·h from 220 panels + 9400kW·h from 94 affected panels).

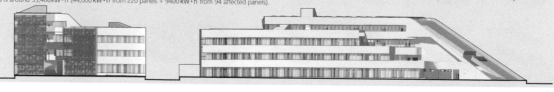

一层平面图 1:300 First Floor Plan

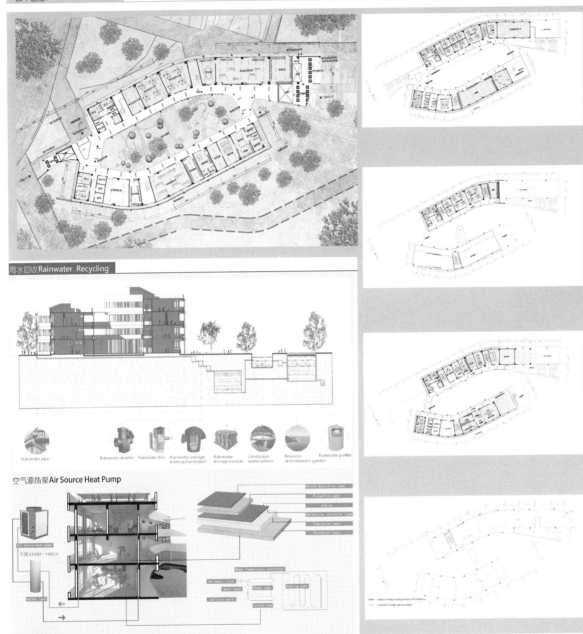

雨水回收 Rainwater Recycling

空气源热泵 Air Source Heat Pump

污水余热回收 Waste Heat Recovery from Sewage

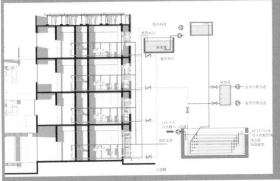

碳排放数据 Carbon Emissions Data

能耗类型	年运行等价电耗（kW·h）	能源形式	能源用量	碳排放因子（tCO₂/单位）	建筑使用寿命（年）	碳排放量（tCO₂）
空调	15951.79	电（kW·h）	15951.79	0.0005703	50.00	454.87
供暖	1236.03	电（kW·h）	1236.03	0.0005703	50.00	35.25
照明	3816.21	电（kW·h）	3816.21	0.0005703	50.00	108.82
电梯	5976.53	电（kW·h）	5976.53	0.0005703	50.00	170.42
生活热水	8621.03	电（kW·h）	8621.03	0.0005703	50.00	245.86
可再生能源	274971.49	电（kW·h）	274971.49	0.0005703	50.00	-7840.81

Indoors, energy-efficient sunlight-simulating task lights are utilized, adjusting based on usage and ambient lighting to maintain a consistent and comfortable workspace. They mimic natural sunlight's color temperature variations, creating a pleasant lighting experience. Coupled with control strategies and wireless motion switches, lighting energy consumption can be reduced by over 65%.

综合奖·入围奖·零碳提升项目
Comprehensive Awards · Nomination · Zero-Carbon Promotion Project

注册号：102006
Register Number：102006

项目名称：竹升·凉岛
Entry Title：Rowing Bamboo, Cool Islands

作者：陈宇航、杨凯、高嘉柔、童彤
Authors：Yuhang Chen, Kai Yang, Jiarou Gao and Tong Tong

作者单位：苏州科技大学
Authors from：Suzhou University of Science and Technology

指导教师：刘长春、金雨蒙
Tutors：Changchun Liu and Yumeng Jin

指导教师单位：苏州科技大学
Tutors from：Suzhou University of Science and Technology

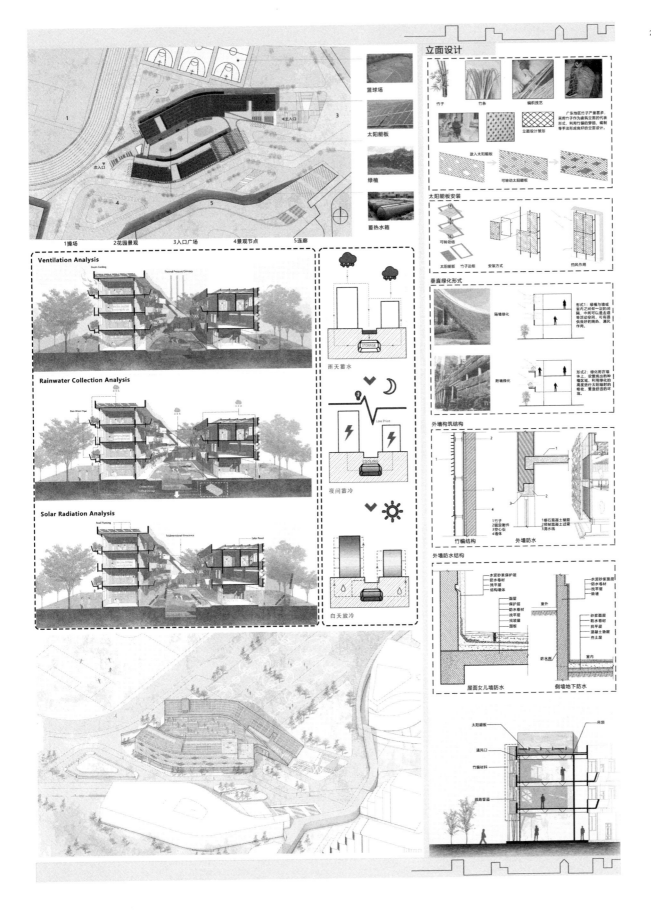

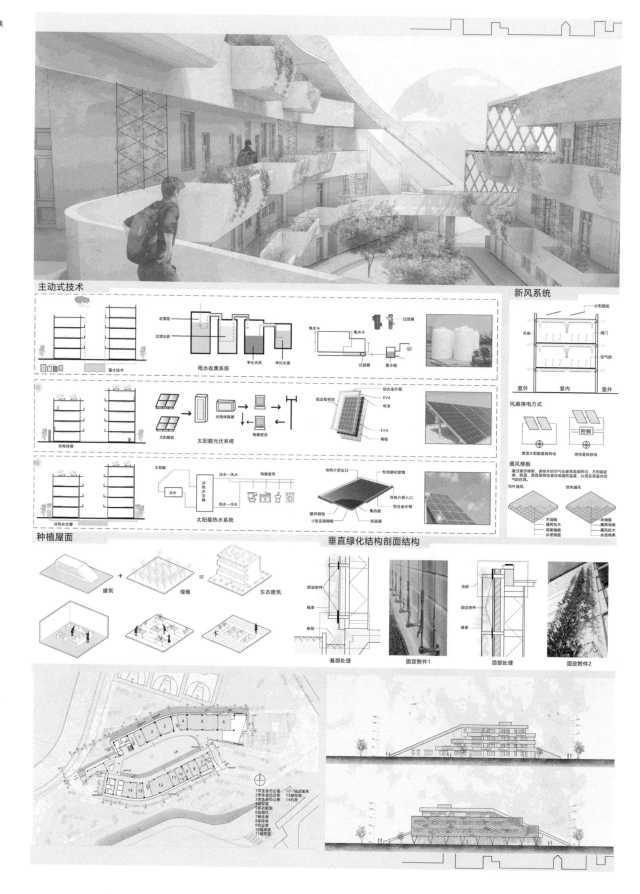

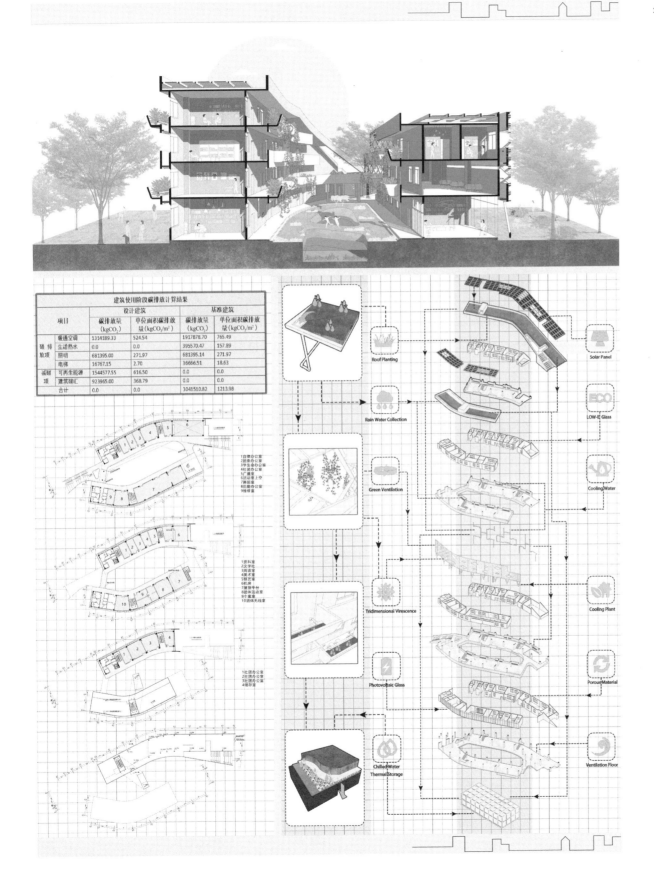

综合奖·入围奖·零碳提升项目
Comprehensive Awards · Nomination · Zero-Carbon Promotion Project

注册号：102007
Register Number：102007

项目名称：光·帆
Entry Title：Light, Sail

作者：李思璇、朱妍、陈珂、刘昊、要敏明
Authors：Sixuan Li, Yan Zhu, Ke Chen, Hao Liu and Minming Yao

作者单位：合肥工业大学
Authors from：Hefei University of Technology

指导教师：王旭
Tutor：Xu Wang

指导教师单位：合肥工业大学
Tutor from：Hefei University of Technology

Site Analysis

Located in the advanced technical school

Surroundings analysis

Climate Analysis

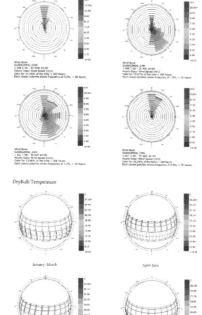

Sunlight analysis

The project is located in the subtropical maritime monsoon climate, with high temperature, abundant rainfall, few frost days and sufficient sunlight. The general climate is characterized by hot, rainy, long summer without winter.

设计说明 /Captions

该改造以低碳为出发点，通过可再生能源的利用、建筑自身的被动节能措施以及具有固碳功能的景观种植设计，实现项目的低碳、零碳目标。改造结合广州气候特点和场地特征，从遮阳、通风、降温三个主要方面进行被动式建筑改造，包括倒退式自遮阳、可调节遮阳板等、应用建筑一体化光伏系统进行主动式节能，同时加入模块化预制屋顶花园、雨水收集系统等方法使建筑降碳节能。场地设计采用水面与绿化阶梯相结合，打造可游走的生态路径，从而实现建筑、环境、人的和谐与平衡发展。

The transformation takes low carbon as the starting point, through the use of renewable energy, the passive energy saving measures of the building itself and the landscape planting design with carbon sequestration function, to achieve the low carbon zero carbon goal of the project. In combination with the climate characteristics and site characteristics of Guangzhou, passive building transformation is carried out from the three main aspects of shading, ventilation and cooling, including regressive self-shading, adjustable sunshade, etc., the application of building integrated photovoltaic system and intelligent control for active energy saving, and the addition of modular prefabricated roof garden, rainwater collection system and other methods to reduce carbon and energy saving of the building. The site design uses the combination of water surface and green stairs to create an ecological path that can be walked, so as to realize the harmonious and balanced development of architecture, environment and people.

Design Strategies

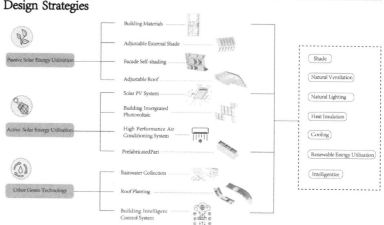

Site Plane 1:400

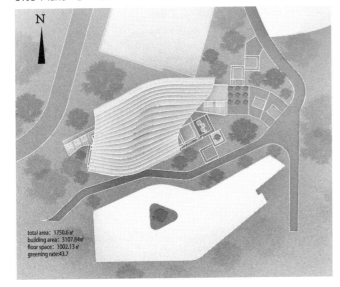

total area: 1750.6 ㎡
building area: 3107.84 ㎡
floor space: 1002.13 ㎡
greening rate: 43.7

Passive Energy Saving Strategy Generation

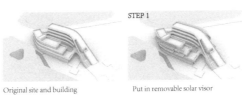

Original site and building | Put in removable solar visor

STEP 2 Into the eco-box | STEP 3 Put in modular roof greenery

STEP 4 Put in an adjustable roof | STEP 5 Site layout drop garden

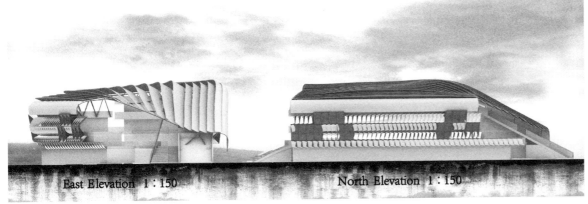

East Elevation 1:150   North Elevation 1:150

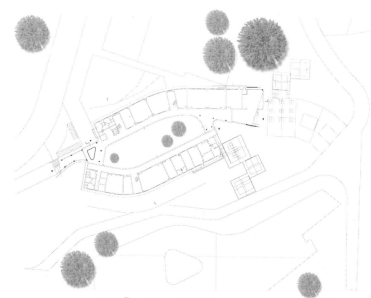

First Floor Plan 1 : 300

Second Plan 1 : 300

Third Plan 1 : 300

Modular Roof Gardem

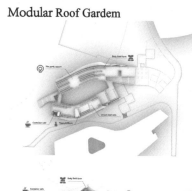

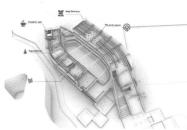

Solar Shading Technology

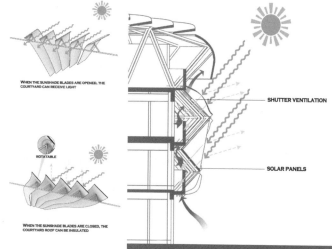

Passive Shading

Natural Ventilation

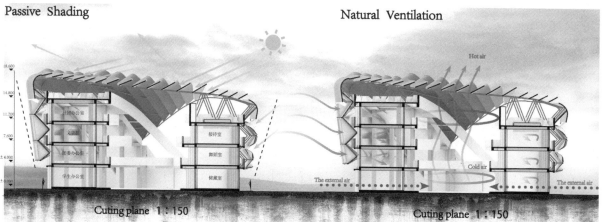

Cuting plane 1 : 150　　　　　　　　　　Cuting plane 1 : 150

Carbon Emission Index Table

project		Calculation results of carbon emissions during the use phase of the building			
		Design buildings		Benchmark building	
		Carbon emissions ($kgCO_2$)	Carbon emissions per unit area ($kgCO_2/m^2$)	Carbon emissions ($kgCO_2$)	Carbon emissions per unit area ($kgCO_2/m^2$)
Carbon emission items	HVAC	146974.22	53.43	213705.85	77.69
	Domestic hot water	674947.40	245.36	674947.40	245.36
	illuminating	0.0	0.0	0.00	0.00
	lift	92322.52	33.56	95613.87	34.76
Carbon reduction items	renewable energy	1118005.96	406.42	0.00	0.00
	Building carbon sinks	199012.95	72.35	0.00	0.00
合计		0.0	0.0	984267.12	357.81

Notes 1 The above calculation results are based on the floor area (public construction)

project	Review of carbon intensity indicators during the use phase of buildings		
	numeric value	Standard requirements	Whether the requirements are met
Reduction in carbon intensity ($kgCO_2/m^2 \cdot a$)	7.16	≥7	satisfy
Carbon intensity reduction rate (%)	100	≥40	satisfy
Review conclusions	Meet the requirements of Article 2.0.3 of the General Code for Building Energy Conservation and Renewable Energy Utilization.		

Build Integrated Photovoltaic(BIPV)   Rainwater Collection System

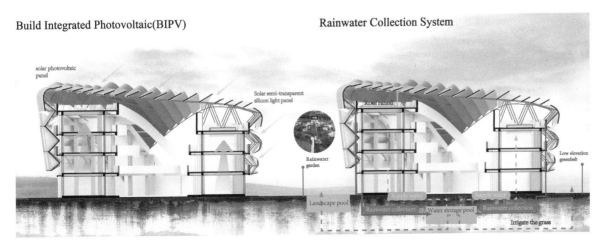

有效作品参赛团队名单
Name List of All Participants Submitting Valid Works

作品编号	作者	单位名称	指导教师	单位名称
101274	王华蕊、朱煜东、王楚云、阙钰霏	长安大学	刘凌	长安大学
101304	田晓可、张昕岩、肖筱侬、赖哲航	厦门大学	贾令堃、石峰	厦门大学
101305	付鑫萍、徐睿瑾、任筱丹、卢婧汐	山东科技大学	冯巍	山东科技大学
101309	汪郑政、高望皓、项涛、赵文祥、武泽宇、翁得玛克·库晚、韩世仓、许文博、吉南木·木沙、乃依马·麦麦提艾则孜	新疆大学	樊辉、艾斯卡尔·模拉克、陈善婷	新疆大学
101313	薛玉炜、郭泓希、周翔宇、王振宇	西南交通大学	王俊	西南交通大学
101316	江芊雨、戚雨田	中国矿业大学	段忠诚、邵泽彪、马全明	中国矿业大学
101320	杨林、卢一迪、刘静文、胡博雅、刘冰	新疆大学、西安建筑科技大学、哈尔滨工业大学、同济大学	王万江、何泉、孟琪	新疆大学、西安建筑科技大学、哈尔滨工业大学
101321	刘为群、孙艺玮	厦门大学	张燕来	厦门大学
101330	牛嘉琪、张燕燕、卢韵莹、倪英杰	中国矿业大学	段忠诚、马全明、邵泽彪	中国矿业大学
101334	李明轩、姬雨杉、张雨桐、刘淑妍、王硕阳、王乐妍	吉林建筑大学	赫双龄、周春艳	吉林建筑大学
101356	罗俊、刘佳、赖胜铭	南宁学院	吴扬	南宁学院
101358	卫茹冰、赵子瑶、陈东晓、孔令熙、王垚橙、舒琨狄	俄罗斯乌拉尔联邦大学、俄罗斯圣彼得堡彼得大帝理工大学、北京建筑大学、华北水利水电大学乌拉尔学院、天津大学	尼基蒂娜·娜塔莉亚·帕夫洛芙娜、塔蒂亚娜·列昂尼多夫娜·西曼金娜、李春青	俄罗斯乌拉尔联邦大学、俄罗斯圣彼得堡彼得大帝理工大学、北京建筑大学
101363	孟圆、孙怡文、陈诗扬、杨一涵、马超毅、陈胤企	长安大学、东北林业大学、阿德莱德大学	夏博、崔鹏	长安大学、东北林业大学
101370	张徐晔阳、朱珂、杨林、余嘉懿、马宏宇	南京工业大学、新疆大学	姜雷、胡占芳	南京工业大学
101371	黄晓、张黔渝、浮英媛	昆明理工大学	谭良斌	昆明理工大学
101373	陈嘉贝、赵玥、李啸跃、李昱瑶	山东建筑大学	侯世荣	山东建筑大学
101385	马明航、赵珺瑶、孟新月、王辰涵	华北理工大学	檀文迪、姚远疆、韦亮	华北理工大学

续表

作品编号	作者	单位名称	指导教师	单位名称
101386	陈宝宇、易柯岑	北京建筑大学	郝石盟	北京建筑大学
101387	姜舒馨、袁孜涵、范劲隆、徐屹东	山东科技大学	冯巍	山东科技大学
101388	韦志绐、杨绍广、于兴、谭小凤、班富猛	南宁学院	吴扬	南宁学院
101392	谢楚湉、慈青松、黄丽文	北京建筑大学、南京工业大学	孙立娜、孙嘉男	北京建筑大学、北京优优星球教育科技有限公司
101396	蔡秀蕴、刘贤、万子豪、黄伟峰、高昕冉、蒋依珊	浙江理工大学	文强	浙江理工大学
101398	王一琳、武昊阅、杨沛东、张琦	南京工业大学	董凌	南京工业大学
101403	杨静怡、郑晨雨、俞俊琛、汤宇豪	福州大学	吴木生	福州大学
101406	卓越、赵钧彦、邓苏媛、阮景添、李梓杰	福州大学	邱文明	福州大学
101408	曾弥纶、翁怡晨、李宸晔、张仪	福州大学	王炜	福州大学
101411	卢植	福州大学	吴木生	福州大学
101419	王重懿、王海婧、蔡飞雨	福州大学	吴木生	福州大学
101420	张康俊、叶以哲、郝雅涵、许馨文	福州大学	崔育新	福州大学
101425	徐艳芳、李莹、方溢凯	厦门大学	贾令堃、石峰	厦门大学
101438	晏肇键、周言婕	北京建筑大学	郝石盟	北京建筑大学
101439	朱诚浩、林锐雄、丁伯绅	福州大学	王炜、吴志刚	福州大学
101440	赵亦佳、李佳欣	北京建筑大学	郝石盟	北京建筑大学
101441	许轩宁、梁中杰、洪宇倩、谭淏蓝	中国矿业大学	段忠诚、姚刚、邵泽彪、马全明	中国矿业大学
101443	马诗淇、陈芝妍、陈佳双豪	福州大学	吴木生	福州大学
101446	杨梦琳、田芳已	北京建筑大学	蒋芳	北京建筑大学
101447	严淑蓝、朴帅同	青岛理工大学	舒珊	青岛理工大学
101451	周凯文、郑泽楠、金函羽、肖雨昕	福州大学	王炜	福州大学

续表

作品编号	作者	单位名称	指导教师	单位名称
101453	顾欣妤、吴欣睿	中国矿业大学	段忠诚、邵泽彪、马全明	中国矿业大学
101457	徐一、艾力克尔·艾百都拉、陈浩	浙江理工大学	邓小军、文强、卜昕翔	浙江理工大学、同济大学建筑设计研究院（集团）有限公司
101463	钟玉萩、程晓华、桑瑞雷	北京交通大学	张文、周艺南	北京交通大学
101467	史梦秋、王翔、赵萍、吕宇翔、李雅宁、刘晓琪、辛瑞铭、来晨莹、陆一帆、张雪珂、刘孟雅	河南大学、黄河科技学院	相恒文、时雪莹、史岩	河南大学、黄河科技学院
101469	刘一卉、张诗崎、白珈铭、程良睿	北京交通大学	张文、周艺南	北京交通大学
101471	李嘉、和莉坤、邱泽龙、张佳琪、亢晨晨、赵莫帆、李明岩、谭雯雯	山西大学	靳维、徐清浩	山西大学
101472	张舒媛、张玉依、蒋林志、梁刚奇、刘国威	南宁学院	吴扬	南宁学院
101475	李雅宁、刘晓琪、辛瑞铭、史梦秋、赵萍、王翔、吕宇翔、司双颖、曹伟毅、闫世纪、张培然	河南大学、黄河科技学院	相恒文、时雪莹、史岩	河南大学、黄河科技学院
101477	马宏宇、周星波、陈何睿、郑明明	南京工业大学	姜雷	南京工业大学
101485	项钰璘、龚佳琳、邱星宇	南京工业大学	刘静萍	南京工业大学
101487	刘超、旦巴曲扎、于豪	西藏大学	索朗白姆	西藏大学
101493	陈天宇、刘祺琪、张智源	福州大学	崔育新	福州大学
101500	马欣瑶、杨欣妍、吴郅诣	上海大学	羊烨	上海大学
101503	席建龙、谢佳良	中国矿业大学	邵泽彪、段忠诚、马全明	中国矿业大学
101506	蔡资潇、贾英杰、陈子墨、文一博	东北大学	刘哲铭	东北大学
101507	岳琮达、胡浪浪、刘嘉琪、申佳梅	沈阳建筑大学	高畅、侯静	沈阳建筑大学
101508	覃莹莹、郭紫玥	河南工业大学	张华、马静	河南工业大学
101515	梁家维、黄凌华	福建农林大学		
101516	黄亮皓、杨鏊	中国矿业大学（北京）	曹颖、贺丽洁	中国矿业大学（北京）
101522	吴颖怡、方瑶璇、杜晟、卢潇月	东北大学	刘哲铭	东北大学

续表

作品编号	作者	单位名称	指导教师	单位名称
101527	战晓琦、宋铭、田子一、温德、张亚雪	北京建筑大学	俞天琦	北京建筑大学
101528	康孟琪、赵琳、吕铭洁、姚望、王君宜	南京工业大学	胡振宇、张海燕	南京工业大学
101530	胡一鸣、姚黄城	福州大学	吴志刚	福州大学
101534	王秉仁、李明远、常昊、梁浩楠、张忠瑞、徐端、刘轩彤	天津大学	郭娟利、李伟	天津大学
101539	王馨可、张月阳	中国矿业大学	马全明、段忠诚、邵泽彪	中国矿业大学
101545	严盟、赖科凤、谢宛彤	广州大学	席明波、万丰登	广州大学
101546	熊昕、陈丁綮	中国矿业大学	邵泽彪	中国矿业大学
101552	丁佳钰、吴欣然、杨栋、易凡、梁红欣	南京工业大学	罗靖、陈建标、刘静萍	南京工业大学
101558	高济美、富诗元、魏君然	沈阳建筑大学	张帆	沈阳建筑大学
101561	王思雨、张越凡、柴静静	河南工业大学	张华、马静	河南工业大学
101563	巩芳芳、蔡宇翔、钟小敏、宗子昌	厦门大学	石峰	厦门大学
101569	钟璐阳、郑诺、阮开沛、祝嘉成、雷远芳	西南交通大学	张樱子	西南交通大学
101570	张彩丽	北京交通大学	田仁青	北京交通大学
101584	陈蓉	湖北工业大学	李竞一	华中科技大学
101595	王珏、邹旭、王艺霖、唐蕾、古阳光、苏陈鑫	南京工业大学	罗靖、陈建标、刘静萍	南京工业大学
101611	叶碧珊、梁振彪、吴树祺、袁鑫桐、郑泽科、蔡锶琦、刘依婷、杨文锦、蒋宇扬	广东工业大学	吉慧、林瀚坤、王树希、董涧清、邓寄豫	广东工业大学、珠海中建兴业绿色建筑设计研究院有限公司、广州市图鉴城市规划勘测设计有限公司
101613	罗嘉仪、许峣崎、姚文星、程麟皓	重庆大学	周铁军	重庆大学
101617	周珺童、宁可、何毅贤、梅颢耀、郭映雪、杨佳成、郝建科、陈佳言	西安交通大学、华南农业大学、香港中文大学	王海旭、徐怡珊、虞志淳	西安交通大学
101623	张旌鹏、李怀宇、刘语嫣	重庆大学	周铁军、张海滨	重庆大学

续表

作品编号	作者	单位名称	指导教师	单位名称
101627	姚湫馨、曾苗威	惠州学院	曾辉鹏	惠州学院
101631	韩杨、杨雨霖	东北石油大学	张永益、李清	东北石油大学
101632	权士博、李昌政、尹硕、郁伟伟	南京工业大学	薛春霖	南京工业大学
101637	龚新宇、马成	中国矿业大学	马全明	中国矿业大学
101638	宋坤阳、刘奕彤、信彦涛、李子涵	沈阳建筑大学	张龙巍、鞠叶欣	沈阳建筑大学
101643	代一卓、张慧、王芯、李少丹、李晓天	广州华立学院、海南省设计研究院有限公司	郭楠、陈秋帆、李超	广州华立学院
101648	田农车夫、魏心怡、戴嘉业、马裕婷、李嘉璐	长安大学	夏博	长安大学
101650	李文钰、李金琦、林永康	南京工业大学	舒欣、吕明扬	南京工业大学
101653	张曦元、马圣新	东南大学	寿焘	东南大学
101658	张欣轶、胡双成、陈伟、周子文	烟台大学	郑斌	烟台大学
101659	高翔、晏广阔、王维昊、宋佳丽、王宇清、段博怀、高辰德、郭振伟、李孟霖、王敬宇	山东建筑大学	郑斐、王月涛	山东建筑大学
101664	殳俊陶、许晓蔓、许可、朱嘉敏、王欣	浙江理工大学	文强	浙江理工大学
101666	许洁治、庄晓茵、翁世峰	惠州学院	曾辉鹏、曾红亮、骆超	惠州学院
101667	王采风、倪鑫鹏、王天缘	南京工业大学	罗靖	南京工业大学
101670	王逸伦、韩芷瑶、周妙菲、徐杨阳	重庆大学	周铁军	重庆大学
101673	李浩阳、胡万杨、骆灿、李雨桐	河南工业大学	毕芳、朱兵司	河南工业大学
101676	崔楠、罗文丰、吴邦奇	沈阳建筑大学	武威	沈阳建筑大学
101679	王晓颖、郝毅卓	烟台大学	郑彬	烟台大学
101680	张锟、张妮、王晓艳、刘甲一、南涵、张一鸣	西北工业大学	邵腾、王晋、刘煜	西北工业大学
101686	黄鑫怡、陈香序、陈思宇、王景琳	北京交通大学	张文、王鑫	北京交通大学
101687	尹涤非、申世峰、王石满、李铭雨	山东建筑大学	何文晶、薛一冰	山东建筑大学

续表

作品编号	作者	单位名称	指导教师	单位名称
101688	巫启隽、李泽清、王祎铭	厦门大学	韩洁	厦门大学
101690	吕萌、张万瑜、胡诗媛	北京交通大学	胡映东、杜晓辉	北京交通大学
101694	赵无极、王汝昊、黄佳鑫、王璐	苏州大学	韩冬辰、郭小平、吴国栋	苏州大学
101702	张学海、曾宇栋	南宁理工学院	傅艺兵	南宁理工学院
101703	梁露丹、陈莹洁	华侨大学	黄鹭红	华侨大学
101708	信蔚林	长安大学	李竞一	华中科技大学
101710	黄泽宇、程雅雯、胡霞胜、缪珂	南京工业大学	舒欣	南京工业大学
101714	赵翔宇、张冉晨、史国崇、庞志祥、张春杰	大连理工大学	李国鹏	大连理工大学
101716	马千里、李睿婧、郑翔书、刘鸿霖、刘玉灏	华南理工大学	赵立华	华南理工大学
101717	毛晓敏、张继元、邓红	西安科技大学	孙倩倩	西安科技大学
101718	徐晓彤、王子旭、高研盛	烟台大学	郑彬	烟台大学
101721	王芊瑶、张皓麒、周小琴、龙思彤、李文苑	广州大学	庞玥、李丽	广州大学
101726	刘意平、毛思怡、马君、王博轩、何兆楠	长安大学	刘凌	长安大学
101730	陈子卿、卜天一、赵依	南京工业大学	南京工业大学	南京工业大学
101731	邢惠玥、石伊、滕震东	南京工业大学	吕明扬、罗靖	南京工业大学
101738	孙艺玮、余娜、林晶晶、刘为群	厦门大学	石峰、孙明宇	厦门大学
101740	王婧一、崔云婷、张佐伊	沈阳建筑大学	侯静	沈阳建筑大学
101741	孙丁超、张啸伊	苏州科技大学	刘长春、陈守恭、金雨蒙	苏州科技大学
101742	李金卓、崔雄飞	北京交通大学	张文、周艺南	北京交通大学
101745	倪彬程、徐小清、赵俊	苏州科技大学、南京工业大学	刘长春、金雨蒙、陈守恭	苏州科技大学
101748	来志慧、浦子弋、张瑜轩、魏祥	南京工业大学	刘静萍	南京工业大学
101749	张雅和	沈阳建筑大学	曹昌浩	杭州嘉树设计咨询有限公司

续表

作品编号	作者	单位名称	指导教师	单位名称
101753	吴奕莹、贺川、颜廷旭	多伦多大学、Farrow Partners、Stepin 设计工作室		
101755	刘家琦、方雪、荆隆庆、赵立朋	石家庄铁道大学	何国青	石家庄铁道大学
101759	蒲海鲲、陈思钰、钟琦	厦门大学	石峰、孙明宇	厦门大学
101761	冯晓潼、靳立越、张珏澜、陈安娴	重庆大学	何宝杰	重庆大学
101763	刘建、刘聪、王爱华、石凯峰、宋中华、李德民	东营市建筑设计研究院、山东省建筑科学研究院有限公司		
101766	杨芳茗、南又源	重庆大学	周铁军、谢崇实、张海滨	重庆大学、重庆设计集团有限公司
101774	蒋辰瑜、彭榆棋、蒙约	重庆大学、东南大学	周铁军、谢崇实、张海滨	重庆大学、重庆设计集团有限公司
101780	马明新、傅廷婷、仲夏仪、周思乔	北京建筑大学	刘烨	北京建筑大学
101781	范可及、曹弋航、俞嘉炜	苏州科技大学	金雨蒙	苏州科技大学
101782	彭硕、杨鑫杰	石家庄铁道大学	高力强	石家庄铁道大学
101783	王亨拓、刘子琦、丁蛟、卫子昂	兰州理工大学	闫幼锋	兰州理工大学
101785	周邑恬、管子淇	河南工业大学	张华、马静	河南工业大学
101789	刘炀、汪益扬、蒋文婷	西安建筑科技大学	李欣、成辉	西安建筑科技大学
101791	李汶哲、洪冬梅、张纵驰、贺羽菲、裴宇骁	西南民族大学	张湮、熊健吾	西南民族大学
101792	张徐聪、邓越辰、汪浩	南京工业大学	董凌	南京工业大学
101795	江超函、张昱萌、佟瑶	北京交通大学	杜晓辉、胡映东	北京交通大学
101799	鞠力、罗思阳、夏湘宜、孔令辉、赖坤锐、胡文通、李建锋	华南理工大学	肖毅强、吕瑶	华南理工大学
101800	徐霄、王耀梅、郑曼如	厦门大学	李芝也、李鍏翰、李渊	厦门大学
101803	贾志彬、孙沐科、崔桐、盖婉婷	西安建筑科技大学	李立敏	西安建筑科技大学
101804	马子雯、曹高源、冯康芸、卞雨馨、许天一	天津大学	朱丽	天津大学
101806	尼文轩、关越泠、李淑媛	南京工业大学	薛洁、刘强	南京工业大学

续表

作品编号	作者	单位名称	指导教师	单位名称
101807	马旮峰、蒋博婧、张同乐、李晓涵、叶孙博	南京工业大学	罗靖、刘静萍、陈建彪	南京工业大学
101809	周圣凡、李明、黄鑫、路嘉林、吴雨荻	长安大学	夏博	长安大学
101810	韩海龙、赵强、翟晨蕾、高涵霖、赵浩辰、徐长印	中建八局第二建设有限公司	徐备、梁汝鸣	中建八局第二建设有限公司
101811	宋郭睿、董春朝、王艺洋、张吉森、李文强、赵鑫慧、张梦婧、张一鸣、贺富年	西北工业大学、中铁十二局数字土木分公司	刘煜、宋戈、曹建	西北工业大学
101822	谢梓怡、张瑄翊、王伟桐、刘宣贝、郑晓琴、杨舒雅	华侨大学	吴正旺	华侨大学
101829	常广皓、吕金书、傅梓豪	南京工业大学	吕明扬	南京工业大学
101831	华鸣凯、李云熙、金晶、郑成、丁方舟	苏州科技大学、上海大学、南京工业大学	陈守恭、刘长春、金雨蒙	苏州科技大学
101839	冯月清、梁中颖	中国矿业大学	马全明、邵泽彪、段忠诚	中国矿业大学
101840	戴洲祺、叶天乐	苏州科技大学	胡炜	苏州科技大学
101841	张小杉	广东培正学院		
101843	林裳、沈鑫、赵雅雯、梁心	内蒙古工业大学	伊若勒泰、许国强	内蒙古工业大学
101844	黄丽文、胡霞胜、缪珂	南京工业大学	刘强	南京工业大学
101845	邓苏媛、李梓杰、阮景添、卓越、赵钧彦	福州大学	邱文明	福州大学
101850	颜学峰、林华淳、李隽奕	华侨大学	施建文	华侨大学
101856	赵胜凯、冯建伦、张育旗、陈鹏宇	兰州交通大学	康华军	兰州交通大学
101858	陈华源、杨键铧、简然、车玉姝、汪珈亦、徐繁	北京建筑大学、云南大学、昆明理工大学、西北工业大学、山东建筑大学	蒋方、吴孟奇、陈虹羽	北京建筑大学、云南大学、昆明理工大学
101860	杜林涛、韩晨阳、毕雪皎、李洁、徐雪健	天津大学	严建伟、杨崴	天津大学
101861	方政、朱长清、丁军行	山东建筑大学	薛一冰	山东建筑大学
101864	朱冰洋、郭宝泽、徐永旺	南京工业大学	薛洁	南京工业大学
101865	郑曼如、王耀梅、徐霄	厦门大学	李渊、李鍏翰、李芝也	厦门大学
101875	李艺菲、王秋雨、张程锦、王玉琦	烟台大学	郑彬	烟台大学
101879	吴浩朴、赵子慧、叶萌、王钰鑫	河南大学	康永基、宗慧宁	河南大学
101884	邹晶晶、谢晨露	中国矿业大学	马全明、段忠诚、邵泽彪	中国矿业大学

续表

作品编号	作者	单位名称	指导教师	单位名称
101886	陈星晨、刘亚洲、高海翔、王国龙	青岛工学院	杨雪娟、赵丹会、陈建伟	青岛工学院
101892	陈梦甜、朱乐奕、谢茜媛	浙江理工大学	文强	浙江理工大学
101896	刘杰新、李明轩、倪浩	南京工业大学	罗靖、张莹莹、徐琦	南京工业大学
101898	李冀、卢艺文	重庆大学	王雪松	重庆大学
101899	罗璘曦、韦海璐、郭家玥、角元昊、王岚、陈文顾、王颖旭	东南大学	刘一歌、王伟	东南大学
101900	向丰裕、延陵思琪、文静、谢乐、宋海波	西南交通大学	张樱子	西南交通大学
101901	李骏安、何子俊、姜春霖、杨姝婷、王奕辰	东南大学	沈宇驰	东南大学
101902	施忱希、刘香帅、史欣瑞	南京工业大学	罗靖、徐琦	南京工业大学
101903	安莹、侯靖轩、胡维杭、张天岳、聂宇璇、张梦特	天津大学	朱丽	天津大学
101907	王君宜、程雅雯	南京工业大学	张海燕、胡振宇	南京工业大学
101910	唐博、彭元新、杨洁	南京工业大学	薛春霖	南京工业大学
101911	方欣然、朱晨昕、洪靖翔	南京工业大学	薛洁	南京工业大学
101917	吴珊珊	厦门大学	韩洁	厦门大学
101921	段博怀、高辰德、郭振伟、李孟霖、王敬宇、高翔、晏广阔、王维昊、宋佳丽	山东建筑大学	郑斐、王月涛	山东建筑大学
101922	高静蔚、黎鑫、龙梦钰	南京工业大学	薛春霖、尤伟	南京工业大学
101926	谢丛朵、叶睿文、郑丹彤、唐国言	苏州大学	孙磊磊、王彪、韩冬辰	苏州大学
101928	刘盈、陈泓池、孔敏	四川农业大学	张丽丽	四川农业大学
101929	杨语心、单芷嫣、张鸿翼	北京交通大学	杜晓辉、胡映东	北京交通大学
101933	朱铃慧、曾詠姿	香港知专设计学院	Wilson	香港知专设计学院
101936	刘依萌、王颢潼、范佳诺、李雨韩	烟台大学	郑彬	烟台大学
101944	康睿博、张钰昆、张文静、刘志豪、范君保、刘艺佳	西安交通大学	王海旭、虞志淳、徐怡珊	西安交通大学
101946	陈宗含、马梓桓、邵竞锐、廖俊裕	广州大学	万丰登、席明波	广州大学
101949	南琨彦、闫鑫、朱圆、张豪文	兰州理工大学	闫佑峰	兰州理工大学

续表

作品编号	作者	单位名称	指导教师	单位名称
101950	黄晓熳、许博媛、陈佳彤、林鸿怿	苏州科技大学	周曦	苏州科技大学
101951	范云江、赵若雅	贵州大学	高明明	贵州大学
101953	吕铭洁、姚望、康孟琪、王君宜、赵琳	南京工业大学	胡振宇、张海燕	南京工业大学
101954	杨梦蝶、冯青青、陈玉莲、程倩	沈阳建筑大学	李辰琦、武威	沈阳建筑大学
101961	吴泽欣、许文静、张欣瑶	南京工业大学	吕明扬	南京工业大学
101962	赵可一、韩柯萍	南京工业大学	吕明扬	南京工业大学
101964	葛铮	北京交通大学	张文、周艺南	北京交通大学
101972	黄泽坤、孙怡恒、吴茗、方心琪、罗翔	河南大学	康永基、宗慧宁	河南大学
101982	王艺达、庄颖楠、杜明洋、张涵	重庆大学	张海滨	重庆大学
101990	张雅馨、张孟埵、梁永琪	山东建筑大学	房涛	山东建筑大学
101992	陶林子、耿心怡、张冰艺、李颐恒、刘浩	燕山大学	李倩、陈云凤	燕山大学
101995	李鸽、徐凯、李卓桐	西安建筑科技大学	何文芳	西安建筑科技大学
101997	徐浩晨、邵子沐、柴西妮、王飞、吕依涵	东南大学	伊若勒泰	内蒙古工业大学
102006	陈宇航、杨凯、高嘉柔、童彤	苏州科技大学	刘长春、金雨蒙	苏州科技大学
102007	李思璇、朱妍、陈珂、刘昊、要敏明	合肥工业大学	王旭	合肥工业大学
102009	高源、吕琰、林雨琦	厦门大学	贾令堃	厦门大学
102012	赵庆卓、郭奕岑、张康睿、陈忠耀	重庆大学	周铁军、张海滨	重庆大学
102013	聂乐仪、王依然、王晗、韦茹钰、唐一文	重庆大学	周铁军、张海滨	重庆大学
102021	刘阳、刘菊影、张金飞、刘晓琪、辛瑞铭	河南大学	康永基、宗慧宁	河南大学
102027	徐振华、范哲人、吴浩、石婧、李悠扬	清华大学、自然营造（北京）建筑设计事务所有限公司、北京交通大学、北京城市学院	徐振华、韩聪	清华大学、北京城市学院
102028	左志宇、刘尔航、孙心	苏州科技大学	张昊雁	苏州科技大学
102034	顾奕成、许斐	苏州科技大学	胡炜	苏州科技大学
102036	高海翔、王国龙、陈星晨、刘亚洲	青岛工学院	杨雪娟、赵丹会、陈建伟	青岛工学院

2023台达杯国际太阳能建筑设计竞赛办法
Guide for the International Solar Building Design Competition 2023

竞赛宗旨：

《绿色低碳发展国民教育体系建设实施方案》中明确要将绿色低碳发展融入校园建设作为目标。在校园建设中优先采用节能减排新技术产品，引导教育系统师生牢固树立绿色低碳发展理念，为实现碳达峰、碳中和目标，奠定坚实的思想和行动基础。本届竞赛以广州科教城文化科技馆和广州市公用事业技师学院社团综合楼为题目，加快推动学校建筑低碳化发展，深入推进可再生能源在学校建设领域的规模化应用。

竞赛主题：阳光·零碳建筑
竞赛题目：
　零碳设计项目：广州科教城文化科技馆
　零碳提升项目：广州市公用事业技师学院社团综合楼
主办单位：国际太阳能学会
　　　　　中国建设科技集团中央研究院
　　　　　中国建筑设计研究院有限公司
承办单位：国家住宅与居住环境工程技术研究中心
支持单位：广州市重点公共建设项目管理中心
冠名单位：台达集团
技术支持：北京天正软件股份有限公司
媒体支持：《世界建筑》
评委会专家：Deo Prasad：澳大利亚科技与工程院院士、澳大利亚勋章获得者、澳大利亚新南威尔士大学教授。
　　　　　杨经文：马来西亚汉沙杨建筑师事务所创始人、2016梁思成建筑奖获得者。
　　　　　Peter Luscuere：荷兰代尔夫特理工大学建筑系教授。
　　　　　崔愷：中国工程院院士、全国工程勘察设计大师、中国建筑设计研究院有限公司总建筑师。

Goal of Competition：

The *Implementation Plan for the Green and Low-Carbon Development in the Construction of National Education System* clearly sets up the goal of implementing the "Green & Low-Carbon Development" principle into campus construction. Aiming to achieve the targets of carbon peaking and carbon neutrality, the adoption of new technologies and products favoring energy-saving and emission-reducing becomes a prior consideration in the overall construction process of the campus projects Accordingly, a firm belief on the "Green & Low-Carbon Development" principle with concrete actions should be introduced to and shared by in the educational circles.

The Competition 2023 wishes to help enhance the ideas of Low-Carbon construction and promote the massive utilization of renewable energy resources in the campus construction, thence two Competition Tasks are designated hereby as following:
1. Cultural and Scientific-technological Museum of Guangzhou Science and Education Park
2. Complex Building for Students Societies of Guangzhou Public Utility Technician College

Competition Theme:

Sunshine & Zero-Carbon Architecture

Competition Tasks

1. Zero-Carbon Design:
Culture and Scientific-technological Museum of Guangzhou Science and Education Park
2. Zero-Carbon Improvement Design:
Complex Building for Students Societies of Guangzhou Public Utility Technician College

Hosts:

International Solar Energy Society (ISES)
Central Research Institute of China Construction Technology Group Co.,Ltd. (CCTC)
China Architecture Design & Research Group (CADG)

王建国：中国工程院院士、教育部高等学校建筑类专业指导委员会主任委员、东南大学建筑学院教授。

庄惟敏：中国工程院院士、全国工程勘察设计大师、梁思成建筑奖获得者、清华大学建筑设计研究院有限公司院长。

陈绍彦：新加坡CPG集团首席创新官、新加坡建筑师注册局主席。

张利：全国工程勘察设计大师、清华大学建筑学院院长。

钱锋：全国工程勘察设计大师，同济大学建筑与城市规划学院教授、博士生导师。

仲继寿：中国建筑设计研究院有限公司总工程师、中国建筑学会健康人居专业委员会和主动式建筑专业委员会主任委员。

黄秋平：华东建筑设计研究总院总建筑师。

冯雅：中国建筑西南设计研究院有限公司顾问总工程师、中国建筑学会建筑热工与节能专业委员会副主任。

张宏：东南大学建筑学院教授、博士生导师。

杨明：华东建筑设计研究院有限公司总建筑师、教授级高级建筑师。

袁烽：同济大学建筑与城市规划学院教授、博士生导师、副院长。

宋晔皓：清华大学建筑学院教授、博士生导师，清华大学建筑学院建筑与技术研究所所长，清华大学建筑设计研究院副总建筑师。

任军：天津大学建筑学院教授、天友建筑设计股份有限公司首席建筑师。

刘恒：中国建筑设计研究院有限公司副总建筑师、绿色建筑设计研究院院长。

组委会成员：由主办单位、承办单位及冠名单位相关人员组成。办事机构设在国家住宅与居住环境工程技术研究中心。

Organizer:

National Engineering Research Center for Human Settlements of China

Supported by:

Management Center of Key Projects Construction for Public Facilities of Guangzhou Municipal Government

Sponsor:

Delta Electronics

Technical Support:

Beijing Tangent Software Co., Ltd.

Media Support:

World Architecture

Experts of the Jury Panel:

Mr. Deo Prasad: Academician of Australian Academy of Technological Sciences and Engineering; Winner of the Order of Australia; professor of University of New South Wales, Sydney, Australia.

Mr. King Mun Yeang: Founder and President of T.R.Hamzah & Yeang Sdn.Bhd. of Malaysia; Winner of Liang Sicheng Architecture Award 2016.

Peter Luscuere: Professor of the Department of Architecture, Technology University Delf, Netherlands.

Mr. Kai Cui: Academician of Chinese Academy of Engineering; Master of National Engineering Survey and Design of China; Chief Architect of China Architecture Design and Research Group Co., Ltd. (CADG)

Mr. Jianguo Wang: Academician of China Academy of Engineering, Director of the Academical Council (Architecture Division) of Ministry of Education, professor of Architecture Department of Southeast University.

设计任务书及专业术语等附件：

附件1：零碳设计项目：广州科教城文化科技馆

附件2：零碳提升项目：广州市公用事业技师学院社团综合楼

附件3：专业术语

奖项设置及奖励形式：

综合奖：

一等奖作品：两个项目分别评审出1个一等奖作品，共计2个，颁发奖杯、证书及人民币100000元奖金（税前）；

二等奖作品：两个项目共评审出4个，颁发奖杯、证书及人民币20000元奖金（税前）；

三等奖作品：两个项目共评审出6个，颁发奖杯、证书及人民币5000元奖金（税前）；

优秀奖作品：两个项目共评审出20个，颁发奖杯、证书及人民币1000元奖金（税前）；

入围奖作品：两个项目共评审出30个，颁发证书。

技术专项奖：名额不限，颁发证书。

设计创意奖：名额不限，颁发证书。

Mr. Weimin Zhuang: Academician of China Academy of Engineering; Master of National Engineering Survey and Design of China; Winner of Liang Sicheng Architecture Prize 2019; President of Architecture Design Research Corporation Limited of Tsinghua University.

Mr. Shaoyan Chen: Chief Innovation Officer of CPG Group of Singapore; Chairman of Architects Registration Bureau of Singapore.

Mr. Li Zhang: Master of National Engineering Survey and Design of China; Dean of Architecture Department of Tsinghua University

Mr. Feng Qian: Master of National Engineering Survey and Design of China; Professor and Doctoral Supervisor of College of Architecture and Urban Planning Tongji University.

Mr. Jishou Zhong: Cheif Enginner of China Architecture Design and Research Group Co., Ltd. (CADG); Chairman of Committee of Healthy Habita of the Architectural Society of China (ASC); Chairman of Committee of Active House (ASC) of the Architectural Society of China (ASC).

Mr. Qiuping Huang: Chief Architect of East China Architecture Design and Research Institute Co., Ltd. (ECADI).

Mr. Ya Feng: Chief Consulting Engineer of China Southwest Architecture Design and Research Institute Co., Ltd.; Deputy Director of Special Committee of Building Thermal and Energy Efficiency of the Architectural Society of China (ASC).

Mr. Hong Zhang: Professor and Doctoral Supervisor of School of Architecture of Southeast University.

Mr. Ming Yang: Chief Architecture of East China Architecture Design and Research Institute Co.,Ltd., Professor-level Senior Engineer.

Mr. Feng Yuan: Doctoral Supervisor and Deputy Dean of College Architecture and Urban Planning of Tongji University.

Mr. Yehao Song: Professor, Doctoral Supervisor and Deputy Dean of School of Architecture of Tsinghua University; Diector of Architecture and Technology Institute of Tsinghua University; Deputy Chief Architect of Architecture Design and Research Institute of Tsinghua University (THAD).

Mr. Jun Ren: Professor of School of Architecture, Tianjin University; Chief Architect of Tenio Group.

Mr. Heng Liu: Deputy Chief Architect of China Architecture Design and Research Group Co., Ltd. (CADG); Director of Green Architecture Design and Research Institute of CADG.

Members of the Organizing Committee:

Composed by selected staff of Hosts, Organizer, and Sponsor, the Organizing Committee renders its daily administration in the office of National Engineering Research Center for Human Settlements in Beijing.

Design Specifications and Glossary (See the Appendix):

Annex 1: Zero-Carbon Design Task: Cultural and Scientific-technological Museum of Guangzhou Science and Education Park

Annex 2: Zero-Carbon Improvement Task: Complex Building for Students Societies of Guangzhou Public Utility Technician College

参赛要求：

1. 欢迎建筑设计院、高等院校、研究机构、研发生产企业等单位，组织专业人员组成竞赛小组参加竞赛。

2. 请参赛者访问 www.isbdc.cn，按照规定步骤填写注册表，提交后会得到唯一的注册号，即为作品编号，一个作品对应一个注册号。提交作品时把注册号标注在每幅作品的左上角，字高 6mm。注册时间 2023 年 3 月 25 日～2023 年 8 月 15 日。

3. 参赛者同意组委会公开刊登、出版、展览、应用其作品。

4. 被编入获奖作品集的作者，应配合组委会，按照出版要求对作品进行相应调整。

注意事项：

1. 参赛作品电子文件须在 2023 年 9 月 15 日前提交组委会，请参赛者访问 www.isbdc.cn，并上传文件，不接受其他递交方式。

2. 作品中不能出现任何与作者信息有关的标记内容，否则将视其为无效作品。

3. 组委会将及时在网上公布入选结果及评比情况，将获奖作品整理出版，并对获奖者予以表彰和奖励。

4. 获奖作品集首次出版后 30 日内，组委会向获奖作品的创作团队赠样书 2 册。

5. 竞赛活动消息发布、竞赛问题解答均可登录竞赛网站查询。

Annex 3: Technical Terms

Award Setting and Reward Form:

First Prize: One First-prize for each of the two tasks, two teams will win the First Prize. Two winner-teams will be granted with the Champion Cups, Award Certificates and RMB100,000 (pre-tax) for each team respectively as cash reward.

Second Prize: two Second-prize for each of the two tasks, four teams will be granted with trophies, certificates, and RMB20,000 (pre-tax) for each team respectively as cash reward.

Third Prize: three Third-prize for each of the two tasks, six teams will be granted with trophies, certificates, and RMB5,000 (pre-tax) for each team respectively as cash reward.

Honorable Mention: 10 prizes for each of the two tasks, 20 teams will be granted with certificates, and RMB1,000 (pre-tax) for each team respectively as cash reward.

Nomination: 15 prizes for each of the two tasks, 30 teams will be granted with certificates.

Award of Technical Excellence : no quota limit, certificates issued.

Award of Design Innovation: no quota limit, certificates issued.

Participants Requirements:

1. Professionals from architecture institutes, colleges, research organizations, producers and corporations are sincerely welcomed to form specialized teams and kindly participate in the Competition.

2. Please visit www.isbdc.cn and fill in the on-line registration form accordingly. A Register Number will then be issued automatically to the applicant after successful submission, and this very number is the sole valid code for each work of the applicant. When submitting the final work, kindly note that the Register Number should be marked clearly at the upper-left corner with fixed height of 6mm. The valid period for registration is from March 25th, 2023 to August 15th, 2023.

3. The Participant(s) (also referred to as "Author(s)", "Applicant(s)" or "Competition Team(s)") acknowledge and hereby authorize the Organizing Committee of the Competition to publish, print, exhibit and utilize/practice their competition works.

4. Authors of the entry works acknowledge and hereby agree to have their works compiled in the book of Awarded Works Collection for Competition 2023 (referred to as "book of Collection") for further edit and/or adjustment with the consideration of the publish requirements under the guidance of the Organizing Committee.

Important Notes:

1. The competition works must be submitted electronically through www.isbdc.cn by September 15th, 2023. No other means of submission is recognized.

2. Any kind of disclosure of the authors' information is strictly prohibited in the filing of the entry works, and the breach of this rule will end up by disqualification of the Competition for the applicants.

所有权及版权声明：

参赛者提交作品之前，请详细阅读以下条款，充分理解并表示同意。

依据中国有关法律法规，凡主动提交作品的"参赛者"或"作者"，主办方认为其已经对所提交的作品版权归属作如下不可撤销声明：

1. 原创声明

参赛作品是参赛者原创作品，未侵犯任何他人的任何专利、著作权、商标权及其他知识产权；该作品未在报刊、网站及其他媒体公开发表，未申请专利或进行版权登记，未参加过其他比赛，未以任何形式进入商业渠道。参赛者保证参赛作品终身不以同一作品形式参加其他的设计比赛或转让给他方。否则，主办单位将取消其参赛、入围与获奖资格，收回奖金、奖品及并保留追究法律责任的权利。

2. 参赛作品知识产权归属

为了更广泛推广竞赛成果，所有参赛作品除作者署名权以外的全部著作权归竞赛承办单位及冠名单位所有，包括但不限于以下方式行使著作权：享有对所属竞赛作品方案进行再设计、生产、销售、展示、出版和宣传的权利；享有自行使用、授权他人使用参赛作品用于实地建设的权利。竞赛主办方对所有参赛作品拥有展示和宣传等权利。其他任何单位和个人（包括参赛者本人）未经授权不得以任何形式对作品转让、复制、转载、传播、摘编、出版、发行、许可使用等。参赛者同意竞赛承办单位及冠名单位在使用参赛作品时将对其作者予以署名，同时对作品将按出版或建设的要求作技术性处理。参赛作品均不退还。

3. 参赛者应对所提交作品的著作权承担责任，凡由于参赛作品而引发的著作权属纠纷均应由作者本人负责。

3. The Organizing Committee should public the appraisal process and results online in a timely manner, and be responsible to publish the book of Collection and issue the awards.

4. The awarded teams will receive two copies of the book of Collection by the Organizing Committee within 30 days since the first publication of the book.

5. All the news and Q&A regarding the competition are released on the website for inquiry.

Declaration on Ownership and Copyright:

Before submitting the competition works, all participants are obliged to read carefully the following clauses, fully understand and agree with the terms set forth.

According to relevant national laws and regulations, the competition Hosts affirm that all the "participants" or "authors", who take part in the competition on their own initiative, accept the following declaration irrecoverably regarding the ownership and copyright of their submitted works.

1. Declaration on Originality

Each piece of entry (works) submitted by the participants is hereof original and without any infringement of third party patents, copyrights, trademarks or any other kinds of intellectual property. It has not been published by any media means including but not limited to newspapers, periodicals and websites. Nor it has been applied for any patents or copyrights registrations , or involved in any other kinds of competitions, or any practice in the commercial markets. All the participants assure their works neither being put in any other competition in the same form or being transferred to third party anytime. In case of any breach of this declaration, the competition Hosts have the rights of canceling the qualification of the participants and withdrawal of the awards if any, and reserve the rights of recourse.

2. The Ownership of Intellectual Property

With the vision of popularizing of the entry works and the competition, the participants hereby agree to relinquish the copyrights of all the works to the competition Organizer and Sponsor, but only to retain their rights of authorship. The Organizer and Sponsor are entitled all the benefits by exercising the copyrights including but not limited to the redesigning, producing, selling, exhibition, publishing and promulgation of the works; and also the benefits by applying the works in field construction projects either for self use or authorizing third-party´s use. The competition Hosts are hereby entitled the rights of displaying and promulgating the works. Without authorization, any organization or individual including the authors of the works are not allowed to transfer, copy, reprint, promulgate, extract, edit, publish, or licensing the works.The participants agree hereby to have their names attached with their works when the Organizer and Sponsor make use of the works, and agree to have their works make necessary revision or adjustment in accordance with the technical requirements for publication and field construction. All the works will not to be returned to the participants after submission.

3. The participants should bear the liability of originality for the entry works. Any disputes stipulated due to the the copyrights ownership should be under full responsibility of the participants.

声明：

1. 参与本次竞赛的活动各方（包括参赛者、评委和组委），即表明已接受上述要求。
2. 本次竞赛的参赛者，须接受评委会的评审决定作为最终竞赛结果。
3. 组委会对竞赛活动具有最终的解释权。
4. 为维护参赛者的合法权益，主办方特提请参赛者对本办法的全部条款、特别是"所有权及版权"声明部分予以充分注意。

国际太阳能建筑设计竞赛组委会
网　址：www.isbdc.cn
组委会联系地址：北京市西城区车公庄大街19号（100044）
联系人：鞠晓磊、张星儿、郑晶茹
联系电话：86-010-88377501、86-010-88377372
电子邮箱：isbdc2021@126.com　QQ交流群：49266054
微信公众号：国际太阳能建筑设计竞赛

Announcement:

1. All the parties involving in the competition including but not limited to the participants, jury members and the Organizing Committee hereby agree to accept the terms of the above Declaration without reserved rights.
2. The participants hereby agree to accept in full the decision of the Jury Panel as a final judgement.
3. The Organizing Committee reserves the rights of final interpretation of this Competition.
4. With the consideration of the rights and interests for the competition participants, the Organizing Committee urges all the participants to carefully check the complete terms of the Competition Requirements, especially the part of "Declaration on the Ownership and Copyrights".

Organizing Committee of Intentional Solar Building Design Competition
Website: www.isbdc.cn
Address: No. 19 Che Gong Zhuang Avenue, Xi Cheng District, Beijing (Postcode 100044)
Liaisons: Xiaolei Ju, Xing'er Zhang, Jingru Zheng
Telephone: 86-010-88377501、86-010-88377372
Email: isbdc2021@126.com　QQ (Chat Group): 49266054
Official Account of WeChat: International Solar Building Construction Competition

附件1：广州科教城文化科技馆项目
Annex 1：Culture and Scientific-technological Museum of Guangzhou Science and Education Park

1. 项目背景

本项目位于广州市增城区广州科教城中轴线以及综合功能区，育水一路与科教大道交叉路口，东经113.67°，北纬23.29°。项目旨在服务整个科教城及城市的需要，为科教城以及城市提供文化科技交流服务功能，形成科教城与城市融合的共享空间；同时充分集成绿色、节能、低碳技术，成为科教城绿色低碳理念的展示窗口。

2. 项目要求

本项目计划建设广州科教城文化科技馆，可有效满足科教城师生的文化、科技方面的需要，同时进行科教城绿色、节能、低碳技术集成的集中展示，实现建筑零碳排放运行。

3. 气候条件

项目所在地为南亚热带海洋性季风气候，气温高、雨量充沛、霜日少、光照充足。多年平均气温为21.6℃，极端高温38.2℃，极端低温-1.9℃，年均降雨量为2039.5毫米。总的气候特点是炎热多雨、长夏无冬。

4. 基础设施

项目场址周边交通较为便利，给水排水、供电等市政配套设施可由场地周边市政道路接入。

5. 竞赛场地

项目所在地主要为空地，总用地面积11447平方米。场地北侧较高，最高处规划标高约36.19米；西侧地势较低，规划标高约为17.47米；周边规划岗水一路标高约为20.76～36.19米、科技大道标高约为17.47～20.76米。场地四周绿化环境良好，植被丰富。场址东北侧规划建设一座110kV变电站。

1. Project Context

The project is located in the central axis of Guangzhou Science and Education City in Zengcheng District, Guangzhou, which is in the integrated functional area, at the intersection of Yushui 1st Road and Science and Education Avenue. It's at 113.67 degrees east and 23.29 degrees north. The project aims to meet the needs of the Science and Education City and Guangzhou, providing cultural and technological exchange services for them, so as to form a shared space for the integration of Science and Education City and Guangzhou. Besides, the project will integrate green, energy-saving and low-carbon technologies and serves as a showcase for the green and low-carbon concept of Science and Education City.

2. Project Requirements

The project of the Culture and Technology Center in Guangzhou Science and Education City must meet the cultural and technological needs of teachers and students in this area, show the integration of green, energy-saving, and low-carbon technology, and achieve zero-carbon emissions in buildings.

3. Climate

The project site has a maritime subtropical monsoon climate characterized by warm and rainy, abundant light and heat, and short frost periods. The average temperature is 21.6℃, with the highest temperature of 38.2℃ and the lowest temperature of -1.9℃. The average annual rainfall is 2039.5 mm. In a word, it is hot and rainy with long summers and no winter.

4. Infrastructure

The project site is surrounded by relatively convenient traffic, water supply and drainage, power supply and other municipal facilities can be accessed by the municipal roads around the site.

5. Basic Information of the Project

The majority of the project site is open space, with a total site area of 11,447 square meters. The northern side of the site is higher; the western side is lower (see the general plan cad drawing for detailed elevations). Surrounding areas have large green areas with rich vegetation. A 110kV substation is planned to be built to the northeast of the site.

图 1 科教城规划总平面图
Figure 1 Site Plan of the Science and Education City

图 2 项目建设范围示意图
Figure 2 Scope of Planning Land

6. 设计要求

1）本项目用地面积约 11447 平方米，项目建设应符合相关规划条件的要求。

2）整体布局应集功能与流线匹配。

3）应考虑当地的气候特点和自然环境，结合当地的建筑材料和建筑特点，合理选用主、被动太阳能技术及适宜的低碳技术，在保证建筑的自然通风、空气质量、采光、用水等需求的同时，降低建筑用能需求和碳排放。

4）将园林场地设计与文化科技等功能有机结合，形成多层次生态景观，体现人文韵味、文化特色和科技创新精神。

5）本项目应采用低碳技术手段实现建筑零碳目标（运行零碳），实现建筑全年的用能与产能平衡，并依据《建筑碳排放计算标准》GB/T 51366 进行建筑运行碳排放计算及减排量核算。

6）应设计建筑碳排放监测系统和建筑智能化控制系统。

7）应考虑低碳材料、可循环材料的使用。

8）考虑项目低碳技术的可展示性、经济性和可推广性。

6. Design Requirements

1) The site area is about 11,447 square meters. The construction should comply with the requirements of the relevant planning conditions.

2) The overall layout should ensure functions and the traffic flow match.

3) Local climate and natural environment should be considered. According to local building materials and architectural features, active and passive solar technologies and low-carbon technologies should be properly applied to reduce building energy usage and carbon emissions while ensuring the building's natural ventilation, air quality, lighting and water usage.

4) The landscape design should be integrated with cultural and technological functions to form a multi-level ecological landscape and reflect the humanistic and cultural characteristics as well as the spirit of technological innovation.

5) The project should adopt low-carbon technologies to achieve the zero-carbon target of the building (operational net zero), strike a balance between energy use and production of the building throughout the year, and carry out carbon emission calculation and emission reduction accounting for the building operation according to GB/T 51366 "Carbon Emission Calculation Standard for Buildings".

6) A building carbon emission monitoring system and an intelligent control system for the building should be designed.

9）竞赛用地及功能设置表如下表所示：

用地及功能设置表

序号	建设内容	建设规模(平方米)	备注
1	文化科技馆用地	11447	
1.1	占地面积	4648	
1.2	绿化	1603	
1.3	道路广场	5196	
2	建筑面积	18000	
2.1	地上建筑面积	14300	
2.1.1	展览教育用房	9100	
2.1.2	公共服务用房	2400	
2.1.3	业务研究用房	1400	
2.1.4	管理保障用房	1400	
2.2	地下建筑面积	3700	只表示出地上出入口即可。需考虑地下停车柱间距

注：室内功能可参照《建筑设计资料集》（第三版）第四分册"科学技术馆"节。

7. 评比办法

1）由组委会审查参赛资格，并确定入围作品。

2）由评委会评选出竞赛获奖作品。

8. 评比标准

1）参赛作品须符合本竞赛"作品要求"的内容。

2）作品应具有原创性，鼓励创新。

3）作品应满足使用功能、绿色低碳、安全健康的要求，建筑技术与太阳能利用技术具有适配性。

4）作品应充分体现太阳能利用技术对降低建筑使用能耗的作用，在经济、技术层面具有可实施性。

评比指标	指标说明	分值
规划与建筑设计	规划布局、建筑空间组合、功能流线组织、建筑艺术	45
太阳能主、被动技术	利用建筑设计与建筑构造实现建筑隔热与通风节能降碳	30
	太阳能光伏、光热等主动太阳能技术的利用实现建筑产能	

7) The use of low-carbon and recyclable materials should be considered.

8) The low-carbon technology should be economic and can be showcased and promoted.

9) The competition site and the function are shown as follows.

Table of Function Settings and Land Use

Number	Construction	Scale (m²)	Note
1	Culture and Technology Center	11447	
1.1	Site Area	4648	
1.2	Green Spaces	1603	
1.3	Roads and plazas	5196	
2	Building Area	18000	
2.1	Ground Area	14300	
2.1.1	Rooms for display and education	9100	
2.1.2	Rooms for Public Services	2400	
2.1.3	Rooms for Business Research	1400	
2.1.4	Rooms for Management and Security	1400	
2.2	Underground Area	3700	Only the above-ground entrances and exits are indicated. The underground garage column spacing should be considered

Note: The interior functions can be found in the section "Science and Technology Museum" in the fourth section of the *Architectural Design Sourcebook* (Third Edition).

7. Appraisal Methods

1) Organizing Committee will check up eligible entries and confirm shortlist entries.

2) Judging Panel will appraise and select out the awarded works.

8. Appraisal Standards

1) The entries must meet the demands of the "Requirements for Works".

2) The entries should be original and innovative.

3) The submitted works should meet the requirements of functions, the green and low-carbon concepts, safety, health, building technology and application of solar energy technology.

4) The work should reflect the role of solar energy technology to reduce the energy consumption of buildings. The technologies can be implemented both at the economic and technical levels.

续表

评比指标	指标说明	分值
采用的其他技术	建造与运行过程中的绿色、低碳、节能等技术	15
可操作性	作品的可实施性，技术的经济性和普适性	10

9. 图纸要求

1）设计深度达到方案设计深度要求，主要技术应有相关的技术图纸和指标。作品图面、文字表达清楚，数据准确。

2）须提交方案设计说明，应包括方案构思、太阳能技术、低碳技术与设计创新（限200字以内），技术经济指标表。

3）提交作品须进行竞赛用地范围内的规划设计，总平面图（含活动场地及环境设计）。

4）提交作品须进行建筑的运行碳排放分析，并能够实现建筑全年碳净零排放。

5）充分表达建筑与室内外环境关系的建筑典型平面图、立面图，比例不小于1：300。

6）能表现出技术与建筑结合的重点部位、局部详图，比例自定相关的技术图、分析图。

7）场地、建筑、局部等效果表现图。

10. 文字要求

1）"建筑方案设计说明"采用中英文双语，其他为英文（建议使用附件3中提供的专业术语）。

2）排版要求：A1展板（594mm×841mm）区域内，统一采用竖向构图，作品张数应为2～4张。

3）中文字体大小于6mm，英文字体不小于4mm。

4）文件分辨率300dpi，格式为JPG或PDF文件。

5）提交参赛者信息表，格式为JPG或PDF文件。

6）上传方式：参赛者通过竞赛网页上传功能将作品递交竞赛组委会，入围作品由组委会统一编辑板眉、出图、制作展板。

Appraisal Indicator	Explanation	Score
Planning and architecture design	Planning concept, layout, traffic streamline organization, architectural art	45
Active and passive solar technologies	Use building design and construction to achieve energy saving and carbon reduction in building insulation and ventilation	30
	Use active solar technologies such as solar photovoltaic and solar thermal to achieve building energy production	
Other technologies	Build and operate the community with green, low-carbon, safe, and healthy technologies	15
Operability	Feasibility, economy, and popularity of works	10

9. Drawing Requirements

1) Entries should meet the project requirement of design depth. Main technologies should have relevant technical drawings and indicators. Drawings and text should be clear and readable with accurate data.

2) Submit a schematic design description containing design concepts, solar energy technology, low-carbon technology, innovative design (less than 200 words), and technical and economic indicators.

3) Provide a planning design within the outline of the competition site and a floor plan (including the venue/environment design).

4) Submissions are required to conduct an analysis of the building's operational carbon emissions and be able to achieve net zero carbon emissions from the building throughout the year.

5) Provide floor plans, elevations, and sections with a scale not less than 1:300, which can fully express the relationship between the architecture and the indoor and outdoor environment.

6) Show the key parts of the combination of technology and architecture and details with other related technical drawings and analysis drawings of self-defined scale.

7) Render perspective drawing of sites, land, and single building.

10. Text Requirements

1) The submission should be in English (technical terms in Annex 3 are recommended), in addition to "architectural schematic design description" in both English and Chinese.

2) Typesetting Requirements: Entries should be put into 2 to 4 exhibition panels, each 594 mm × 841 mm (A1 format) in size (arranged vertically).

3) Word height of Chinese is not less than 6mm and that of English is not less than 4mm.

4) File resolution: 300 dpi in JPG or PDF format.

5) Information tables of participants should also be submitted in JPG or PDF format.

6) Uploading: Entries should be submitted to the organizing committee through the competition's official website. Shortlist works will be compiled, printed, and made into exhibition panels by the organizing committee.

附件2：广州市公用事业技师学院社团综合楼项目

Annex 2：Complex Building for Students Societies of Guangzhou Public Utility Technician College

1. 项目背景

本项目位于广州市增城区广州科教城一期核心地带西侧的广州公用事业高级技师学院内，东经113.67°，北纬23.29°。项目北、西邻近运动场、南邻南食堂，东邻校园主干道。目前项目主体结构已封顶，部分室内隔墙砌筑，暖通、电气、给水排水工程尚未施工。

2. 项目要求

本项目为广州市公用事业技师学院社团综合楼，其主要功能为：学生社团活动、办公用房、信息中心用房等。要求基于现有已封顶的主体结构，不改变现有建筑面积、外形尺寸，采用多种实现建筑节能的建筑构配件和系统设备，同时集成适宜的可再生能源系统，有效实现建筑的自然通风、采光和隔热，降低建筑用能和碳排放，同时利用可再生能源系统覆盖建筑的用能需求，实现建筑全年的用能与产能平衡，建筑运行零碳排放。

3. 气候条件

项目所在地为南亚热带海洋性季风气候，气温高、雨量充沛、霜日少、光照充足。多年平均气温为21.6℃，极端高温38.2℃，极端低温-1.9℃，年均降雨量为2039.5毫米。总的气候特点是炎热多雨、长夏无冬。

4. 基础设施

项目场址周边交通较为便利，给水排水、供电等市政配套设施齐全。

5. 项目基本信息

项目位于广州市公用事业技师学院内，3150平方米（均为地上建筑面积）。地上四层，建筑高度：15.455平方米（屋顶面层），钢筋混凝土框架结构，建筑其他信息详见附件CAD图。项目原设计为绿色建筑二星级标准。

1. Project Context

The project is located in Guangzhou Public Utility Senior Technical School in the west of the core area of Guangzhou Science and Education City Phase I, Zengcheng District, Guangzhou, 113.67° E, 23.29° N. The project is adjacent to the sports field in the north and west, the south cafeteria in the south, and the main road of the campus in the east. At present, the main structure of the project has been roofed. Some indoor partition walls have been built, and the HVAC, electricity, water supply and drainage have not yet been constructed.

2. Project Requirement

The project is Guangzhou Public Utility Senior Technician College Club Complex with the main functions of rooms for student club activities, an office and an information center. The designer must use existing roofed main structure. The building area, height and structure of this project cannot be changed. Adopt a variety of building components and system equipment and integrate appropriate renewable energy to ensure the building's natural ventilation, lighting, and heat insulation and reduce the energy consumption and carbon emission of the building. The design should also use renewable energy systems to cover the building's energy demand, achieving a balance of building energy use and production throughout the year, and zero carbon emissions from building operation.

3. Climate

The project site has a maritime subtropical monsoon climate characterized by warm and rainy, abundant light and heat, and short frost periods. The average temperature is 21.6℃, with the highest temperature of 38.2℃ and the lowest temperature of -1.9℃. The average annual rainfall is 2039.5 mm. In a word, it is hot and rainy with long summers and no winter.

4. Infrastructure

The project site is surrounded by good traffic and equipped with water supply and drainage, power supply and other municipal facilities.

5. Basic Information of the Project

The project is located in Guangzhou Public Utility Senior Technician School with 3150 square meters (all ground area). It is a reinforced concrete frame structure with four floors above ground. Building height: 15.455M (roof surface). See CAD drawings in the annex for other information of the building. The original design of the project is in line with the green building two-star standard.

图 1 项目所在区位图
Figure 1　Project Location Map

图 2 项目鸟瞰图
Figure 2　Vertical View of the Project Site

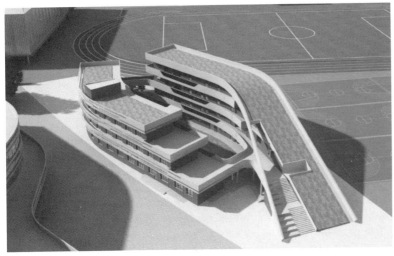

图 3 项目原设计效果图
Figure 3　Original Design Rendering of the Project

图 4 项目施工现状图
Figure 4　Project Construction Status Map

6. 设计要求

1) 本项目建筑面积、建筑高度、建筑结构等均不能改动。

2) 建筑内部功能不宜改动。

3) 应考虑当地的气候特点和自然环境,结合校园整体风貌开展改造设计。

4) 采用多种实现建筑节能的建筑构配件和系统设备,在保证建筑的自然通风、空气质量、采光、用水等需求的同时,降低建筑用能需求和碳排放。

5) 应集成太阳能等可再生能源系统,实现建筑全年的用能与产能平衡。

6) 本项目应采用低碳技术手段实现建筑零碳目标(运行零碳),并依据《建筑碳排放计算标准》GB/T 51366 进行建筑运行碳排放计算及减排量核算。

7) 应用适宜的智能监测、控制系统实现对建筑机电系统的智能控制和碳排放数据监测。

8) 应考虑低碳材料、可循环材料的使用。

9) 考虑项目低碳技术的可展示性、经济性和可推广性。

7. 评比办法

1) 由组委会审查参赛资格,并确定入围作品。

2) 由评委会评选出竞赛获奖作品。

8. 评比标准

1) 参赛作品须符合本竞赛"作品要求"的内容。

2) 作品应具有原创性,鼓励创新。

3) 作品应满足使用功能、绿色低碳要求,建筑技术具有适配性。

评比指标	指标说明	分值
规划与建筑设计	建筑立面及形体设计,建筑气候适应性设计	25
建筑降碳设计	被动节能设计、建筑围护结构节能设计、系统设备节能设计	25
建筑产能设计	太阳能等可再生能源建筑集成设计	20
其他适宜技术	建造与运行过程中的绿色、低碳、海绵城市等技术	10
建筑碳排放计算	建筑碳排放计算、碳减排量核算	10
可操作性	作品的可实施性,技术的经济性和普适性	10

6. Design Requirements

1) The building area, height and structure of this project cannot be changed.

2) The internal functions of the building should not be changed.

3) Local climate and natural environment should be considered. The design of the campus renovation should take its overall appearance into account.

4) Adopt a variety of building components and system equipment to achieve building energy efficiency while ensuring the building's natural ventilation, air quality, lighting and water usage.

5) Renewable energy systems such as solar energy should be integrated to achieve a balance between energy use and production of the building throughout the year.

6) The project should adopt low-carbon technologies to achieve the zero-carbon target of the building (operational net zero) and carry out carbon emission calculation and emission reduction accounting for the building operation according to GB/T 51366 *"Carbon Emission Calculation Standard for Buildings"*.

7) Apply appropriate intelligent monitoring and control systems to achieve intelligent control of building mechanical and electrical systems and carbon emission data monitoring

8) The use of low-carbon and recyclable materials should be considered.

9) The low-carbon technology should be economic and can be showcased and promoted.

7. Appraisal Methods

1) Organizing Committee will check up eligible entries and confirm shortlist entries.

2) Judging Panel will appraise and select out the awarded works.

8. Appraisal Standards

1) The entries must meet the demands of the "Requirements for Works".

2) The entries should be original and innovative.

3) The submitted works should meet the requirements of functions, the green and low-carbon concepts. Building technology should adaptable.

Appraisal Indicator	Explanation	Score
Planning and architecture design	Planning concept, layout, traffic streamline organization, architectural art	25
Building Carbon Reduction Design	Passive energy-saving design, building envelope energy-saving design, system and equipment energy-saving design	25
Building energy production design	Integrated design of solar and other renewable energy buildings	20
Other technologies	Green, low-carbon and sponge city technologies in the construction and operation process	10

续表

Appraisal Indicator	Explanation	Score
Calculation of building carbon emissions	Calculation of building carbon emissions and accounting for carbon emission reductions	10
Operability	Feasibility, economy, and popularity of works	10

9. 图纸要求

1）设计深度达到方案设计深度要求，主要技术应有相关的技术图纸和指标。作品图面、文字表达清楚，数据准确。

2）须提交方案设计说明，应包括方案构思、低碳技术与设计创新（限200字以内），建筑碳排放指标表。

3）提交作品须包含竞赛项目的总平面图（含周边环境）。

4）提交作品须进行建筑的运行碳排放分析，并能够实现建筑全年碳净零排放。

5）充分表达建筑与室内外环境关系的建筑典型平面图、立面图，比例不小于1：300。

6）能表现出技术与建筑结合的重点部位、局部详图，比例自定，相关的技术图、碳排放数据等分析图。

7）场地、建筑、局部等效果表现图。

10. 文字要求

1）"建筑方案设计说明"采用中英文双语，其他为英文（建议使用附件3中提供的专业术语）。

2）排版要求：A1展板（594mm×841mm）区域内，统一采用竖向构图，作品张数应为2～4张。

3）中文字体大小于6mm，英文字体不小于4mm。

4）文件分辨率300dpi，格式为JPG或PDF文件。

5）提交参赛者信息表，格式为JPG或PDF文件。

6）上传方式：参赛者通过竞赛网页上传功能将作品递交竞赛组委会，入围作品由组委会统一编辑板眉、出图、制作展板。

9. Drawing Requirements

1) Entries should meet the project requirement of design depth. Main technologies should have relevant technical drawings and indicators. Drawings and text should be clear and readable with accurate data.

2) Submit a schematic design description containing design concepts, solar energy technology, low-carbon technology, innovative design (less than 200 words), and technical and economic indicators.

3) Provide a planning design within the outline of the competition site and a floor plan (including the venue/environment design).

4) Submissions are required to conduct an analysis of the building's operational carbon emissions and be able to achieve net zero carbon emissions from the building throughout the year.

5) Provide floor plans, elevations, and sections with a scale not less than 1:300, which can fully express the relationship between the architecture and the indoor and outdoor environment.

6) Show the key parts of the combination of technology and architecture and details with other related technical drawings and analysis drawings of self-defined scale.

7) Render perspective drawing of sites, land, and single building.

10. Text Requirements

1) The submission should be in English (technical terms in Annex 3 are recommended), in addition to "architectural schematic design description" in both English and Chinese.

2) Typesetting Requirements: Entries should be put into 2 to 4 exhibition panels, each 594 mm × 841 mm (A1 format) in size (arranged vertically).

3) Word height of Chinese is not less than 6mm and that of English is not less than 4mm.

4) File resolution: 300 dpi in JPG or PDF format.

5) Information tables of participants should also be submitted in JPG or PDF format.

6) Uploading: Entries should be submitted to the organizing committee through the competition's official website. Shortlist works will be compiled, printed, and made into exhibition panels by the organizing committee.

附件3：专业术语
Annex 3：Technical Terms

百叶通风	— shutter ventilation	光伏发电系统	— photovoltaic system
保温	— thermal insulation	光伏幕墙	— PV façade
被动太阳能利用	— passive solar energy utilization	回流系统	— drainback system
敞开系统	— open system	回收年限	— payback time
除湿系统	— dehumidification system	集热器瞬时效率	— instantaneous collector efficiency
储热器	— thermal storage	集热器阵列	— collector array
储水量	— water storage capacity	集中供暖	— central heating
穿堂风	— through-draught	间接系统	— indirect system
窗墙面积比	— area ratio of window to wall	建筑节能率	— building energy saving rate
次入口	— secondary entrance	建筑密度	— building density
导热系数	— thermal conductivity	建筑面积	— building area
低能耗	— lower energy consumption	建筑物耗热量指标	— index of building heat loss
低温热水地板辐射供暖	— low temperature hot water floor radiant heating	节能措施	— energy saving method
地板辐射采暖	— floor panel heating	节能量	— quantity of energy saving
地面层	— ground layer	紧凑式太阳热水器	— close-coupled solar water heater
额定工作压力	— nominal working pressure	经济分析	— economic analysis
防潮层	— wetproof layer	卷帘外遮阳系统	— roller shutter sun shading system
防冻	— freeze protection	空气集热器	— air collector
防水层	— waterproof layer	空气质量检测	— air quality test (AQT)
分户热计量	— household-based heat metering	立体绿化	— tridimensional virescence
分离式系统	— remote storage system	绿地率	— greening rate
风速分布	— wind speed distribution	毛细管辐射	— capillary radiation
封闭系统	— closed system	木工修理室	— repairing room for woodworker
辅助热源	— auxiliary thermal source	耐用指标	— permanent index
辅助入口	— accessory entrance	能量储存和回收系统	— energy storage & heat recovery system
隔热层	— heat insulating layer	平屋面	— plane roof
隔热窗户	— heat insulation window	坡屋面	— sloping roof
跟踪集热器	— tracking collector	强制循环系统	— forced circulation system

热泵供暖	heat pump heat supply	填充层	fill up layer
热量计量装置	heat metering device	通风模拟	ventilation simulation
热稳定性	thermal stability	外窗隔热系统	external windows insulation system
热效率曲线	thermal efficiency curve	温差控制器	differential temperature controller
热压	thermal pressure	屋顶植被	roof planting
人工湿地效应	artificial marsh effect	屋面隔热系统	roof insulation system
日照标准	insolation standard	相变材料	phase change material (PCM)
容积率	floor area ratio	相变太阳能系统	phase change solar system
三联供	triple co-generation	相变蓄热	phase change thermal storage
设计使用年限	design working life	蓄热特性	thermal storage characteristic
使用面积	usable area	雨水收集	rain water collection
室内舒适度	indoor comfort level	运动场地	schoolyard
双层幕墙	double façade building	遮阳系数	sunshading coefficient
太阳方位角	solar azimuth	直接系统	direct system
太阳房	solar house	值班室	duty room
太阳辐射热	solar radiant heat	智能建筑控制系统	building intelligent control system
太阳辐射热吸收系数	absorptance for solar radiation	中庭采光	atrium lighting
太阳高度角	solar altitude	主入口	main entrance
太阳能保证率	solar fraction	贮热水箱	heat storage tank
太阳能带辅助热源系统	solar plus supplementary system	准备室	preparation room
太阳能电池	solar cell	准稳态	quasi-steady state
太阳能集热器	solar collector	自然通风	natural ventilation
太阳能驱动吸附式制冷	solar driven desiccant evaporative cooling	自然循环系统	natural circulation system
太阳能驱动吸收式制冷	solar driven absorption cooling	自行车棚	bike parking
太阳能热水器	solar water heating		
太阳能烟囱	solar chimney		
太阳能预热系统	solar preheat system		
太阳墙	solar wall		